普通高等教育农业农村部"十三五"规划教材
全国高等农林院校"十三五"规划教材

中国农业文化概览

An Overview of Chinese Farming Culture

王 珍 主编

中国农业出版社
北 京

编写人员

主　编　王　珍
副主编　杨红英　刘培昕　杜　敏　符娟娟
编　者　吕巧宁　严立梅　任秋兰　李　梅
　　　　　梁莉莉　王丽慧

An Overview of Chinese Farming Culture

前言 Preface

  中国文化"走出去"战略是我国建设社会主义文化强国的重要举措。随着文化"走出去"步伐的加快，中国优秀传统文化更需要对外传播，这是中国文化发展的需要。中国的农业文化，尤其是古代农耕文明是我国传统文化最核心的组成部分，是中国文化的精华所在。担负着"用英语讲好中国故事、传播好中国声音"这一伟大使命的当代大学生需要了解中国传统农业文化的相关知识、熟知如何用英语表达以及掌握文化阐释的能力。

  中国文化"走出去"的战略背景下，中国文化的英语教材编写也成为高校英语教材建设的重要组成部分。新一代大学生肩负着中国文化"走出去"的重要使命，这是我们编写《中国农业文化概览》的初心。众所周知，当前大学英语教材中普遍存在"中国文化失语"现象，并且大学生母语文化的英语表达能力也比较薄弱。另外，《大学英语教学指南》也明确提出大学英语课程的培养目标之一就是增强学生的跨文化交际能力，同时提高综合文化素养。基于上述三点原因的考虑，《中国农业文化概览》这本英语教材应运而生。本教材编写的最终目的是帮助大学生了解中国传统农耕文化并且能够用英语进行对外交流，提升文化自信。

  本教材的编写指导思想就是依托中国农业文化中最具国际影响力和最典型、最核心的农耕文化思想或现象，循序渐进地引导学生从了解农耕文化到熟知英语表达，再到引发对传统文化现象的思考，最终达到用辩证的观点去看待传统文化现象的目标。基于此，本教材选取了中国农业文化中最具代表性的十个主题作为编写的主要内容，对中国农耕文明的起源、相关的神话传说、农耕思想、传统节日、丝绸之路、瓷器的发展历程还有和我们日常生活密切相关的酒文化和茶文化等话题进行了介绍。同时，本教材在每个单元后面都增设了练习题，包括"判断正误""填空""术语解析""传统文化段落翻译"与"批判性思考"五种题型。练习题的设计遵循了由易到难、由了解认知到辩证思考的编写原则。本教材在选取具体的内容方面充分考虑到了学生的兴趣，比如有关神农氏的传说、具有中国农业百科全书称号的《齐民要术》、古代的丝绸之路、古代庙会的现代演变、古代婚礼的三书六聘、喝酒与饮茶的礼节以及二十四节气的由来等。这些内容可以激发学生对用英语学习农耕文化的内在动机，帮助学生更好地用英语讲好中国的农耕文化故事。

  本教材由青岛农业大学一线教师编写，编写过程中倾注了编者大量的心血，有的还融入了独到的见解和心得。编写组的老师多次召开编写会，反复讨论，几易其稿，最终形成现有教材。本教材既是主编的研究成果，亦是编写团队集体智慧的结晶。本教材在编写的过程中还有幸得到了青岛农业大学外国语学院纪卫宁教授的恳切指导，在此深表谢意。由于编者能力有限，教材难免会有些许错误，也恳请使用教材的教师和学生给予宝贵意见。

编　者

2019年12月

目录 Contents

前言

Chapter 1　Myth and Sacrifice ·· 1
　Section A　Myth ··· 1
　Section B　Sacrifice ·· 12
Chapter 2　Wisdom and Thoughts of Farming ······························ 24
　Section A　Agricultural Policies and Reforms ··························· 24
　Section B　Ancient Agricultural Books ··································· 38
Chapter 3　Twenty-Four Solar Terms ·· 53
　Section A　General Introduction to Twenty-Four Solar Terms ········ 53
　Section B　Farming Activities and Twenty-Four Solar Terms ········· 62
　Section C　Dietary Customs and Twenty-Four Solar Terms ··········· 71
Chapter 4　Typical Chinese Food Crops and Farming Tools ··········· 81
　Section A　Typical Food Crops in China ································· 81
　Section B　Traditional Chinese Farming tools ·························· 89
Chapter 5　Flood Control and Irrigation ···································· 98
　Section A　Yu Taming the Flood ·· 98
　Section B　Dujiangyan Irrigating System ······························· 104
　Section C　Harnessing of the Yellow River ····························· 113
Chapter 6　Traditional Festivals and Folk Customs ······················ 121
　Section A　Traditional Festivals ·· 121
　Section B　Marital and Funeral Customs ································ 131
　Section C　Temple Fair ·· 140
Chapter 7　Silk and Ceramics Culture ······································ 148
　Section A　Culture of Silk ·· 148

中国农业文化概览
An Overview of Chinese Farming Culture

    Section B   Culture of Ceramics ··· 157
    Section C   The Silk Road ··· 172
    Section D   The Belt and Road Initiative ····································· 185
Chapter 8   Chinese Tea Culture ··· 194
    Section A   Origin and Development of Tea ································ 194
    Section B   Tea Varieties and Benefits ·· 205
    Section C   Tea Customs and Etiquette ······································· 214
Chapter 9   Chinese Alcohol Culture ·· 223
    Section A   Origin and Development of Chinese Alcohol ············ 223
    Section B   Drinking Customs ·· 233
    Section C   Alcohol and Social Life ··· 241
Chapter 10   Chinese Food Culture ··· 251
    Section A   Food and Regions ·· 251
    Section B   Food and Festivals ··· 262

课后练习题答案 ·· 273

参考文献 ··· 301

Chapter 1
Myth and Sacrifice

Section A Myth

In every language, the word "myth" describes a popular story accepted as truth or half-truth. Historically, however, myths constitute a body of **speculation** that reflects the collective beliefs of a society. They **rationalize** and explain the unknown, and usually ground those stories on the evidence of the senses. Those diverse myths have mirrored human efforts to explore the unknown, challenge the great dangers of the environment, achieve communal cooperation, and understand the destiny of mankind.

Classic of Mountains and Seas

Most of Chinese myths can find their source in some age-old books, like **Classic of Mountains and Seas**, **The Songs of Chu**, **Tale of King Mu of Zhou**, **Lvshi Chunqiu**, and **Huainanzi**. Some of them have been passed down to even today in the form of picture books, cartoons, movies and video games, such as *Pangu Separates Heaven and Earth*, *Nvwa Makes Men*, and *Chang'e Ascends to the Moon*.

Pangu Separates Heaven and Earth

Almost every culture has produced a creation myth (创世神话). China is no exception. Pangu, Chinese Adam called by Westerners, is said to have created the world by separating heaven and earth.

A version about creation myth in China is that the universe was originally a blurred and chaotic **entity** like an enormous egg. Inside the total darkness of the egg, there slept a giant, Pangu. He has been sleeping for 18,000 years till one day he opened his eyes and found himself confined in a sticky, chaotic and invisible space. He couldn't tolerate the messy darkness, so he **punched** both arms upward against the overhead eggshell. The egg suddenly cracked, split with a bang and swirled dazzlingly. Light and transparent substances from inside **ascended** and turned into heaven while those heavy and **murky** things gradually **deposited** a vast earth. Pangu stood in between, holding up the sky in case heaven and earth might meet again. Every day the sky rose one *zhang* （丈, a unit of length) higher, the earth became one *zhang* thicker then Pangu grew one *zhang* taller. Tens of thousands of years passed, the heaven and the earth stretched beyond measurement and then stopped changing. The world finally came into being.

Pangu

Million years of sky-holding task has exhausted the solitary giant; and thus his life came to an end. The moment of his collapse witnessed a miracle: the air he breathed turned into the wind and the cloud, his left eye into the sun, right eye into the moon, his limbs into the four poles of the earth, his body into the **Five Famous Mountains**, his blood into rivers, his veins into roads, his muscles into fields and his hair into stars in the sky. And the fine hair on his body changed into flowers, grasses, bushes and trees, and his sweat into rain. Ever since then, there appeared sun shining, wind blowing, river flowing and bird singing in the world. Then there walked human on the earth.

After his death, people set up a tomb covering 300 miles in honor of his matchless feat and worship him as the creator （创世神) in many places in China.

Shennong

Shennong, Divine Farmer or Agriculture God, is a mythical ruler in prehistoric China. Some 4,500 years ago, Rensi（任姒）, a girl in today's Shanxi Province, became pregnant after she stared at a red dragon out of the Jiangshui River（姜水）. On the 26th day of the fourth month, she gave birth to an ox-headed, sharp-horned boy, Shennong, who was later chosen as the head of his tribe because of his unchallenged bravery. He made a lifelong commitment to improving the living conditions for his people. At the time, people were subjected

Shennong

to illness and starvation. He crossed mountains and rivers in search of productive crops to secure sufficient food supply. He also invented some tools such as *lei*（耒）, a bent-wood plow, and *si*（耜）, a cut-wood rake to ease the farming burden and increase working efficiency.

Legend holds that Shennong had a transparent belly, and thus it could clearly demonstrate the effects of different herbal substances he ate. Such physical advantage enabled him to identify the **properties** and functions of different herbs easily. The earliest Chinese phar-macopoeia（药典）was entitled **Shennong Herbal Medicine**, which covered more than 300 kinds of medicines from minerals, plants, and animals. Tea, an **antidote** against the **poisonous** effect caused by 70 herbs, was said to be one of his discoveries. The technique of **acupuncture**[1] was also believed

Bent-wood Plow

to be introduced and improved by Shennong. What he has done in medicine won the Father of Chinese Medicine for him.

He was credited with other inventions including making axes, digging wells, practising agricultural irrigation, preserving seeds by using boiled horse urine (尿液), and creating Chinese calendar. And he also **refined** the knowledge of pulse-taking measurements and **moxibustion**[2] and **routinized** the harvesting ceremony. Shennong is recognized as the **deified patron** especially for farmers, rice traders and practitioners of traditional Chinese medicine.

Shennong was **allegedly** poisoned to death when he tried *gelsemium elegans* (断肠草) and was buried in today's Yanling County, Hunan Province. Till today, numerous sacrificial rites are held in different places across China every year to his memory.

Quafu Chases After the Sun

The legendary figure Quafu was the leader of Quafu tribe who settled in a mountain called *Chengdu Zaitian* (成都载天) in the northern wilderness of China. It was said that people in the tribe were known for their **unrivaled** power and extraordinary running ability. These physical advantages made it possible for them to defend the weak against injustice.

One version of the myth *Quafu Chases After the Sun* was that there broke out a widespread drought in about 4,000 BC. The burning sun cracked the land, dried rivers and **scorched** the crops. People searched everywhere for water, but in vain. So Quafu was **indignant** at the sun and swore to snatch it down. He set off on the journey to start a face-to-face fight with the sun. The sun saw through his attempts, then gave off beams of light to stop him. However, Quafu was not intimidated into slowing down. The chase was almost over when Quafu approached the sun's resting place Yugu (禺谷). Nowhere to

Quafu

escape, the sun turned around and poured all its lights upon Quafu, **enveloping** him with extreme heat. Suddenly Quafu became faint with thirst. He planned to catch the sun after drinking water, but was exhausted before reaching the river.

Another version seemed to go to the opposite. Quafu had been thinking about talking the sun into shedding light evenly in four seasons to secure a good harvest for his people. One day he took a stick and set out to catch up with the sun. At that moment the sun was leisurely sitting in the dragon-drawn carriage, then it suddenly caught sight of a giant rushing toward it. Startled, it drove its carriage westward in a haste. While on the ground, Quafu strained himself to run, but failed to narrow down their distance.

Quafu ran after the sun westward to its resting place. Irritated, the sun emitted the beams of sunlight, **engulfed** him and **rendered** him blind. Numerous rivers failed to ease his thirst or relieve his **fatigue**. Quafu collapsed with a sigh. His stick flied to the air and then landed on the ground where **leafy** trees grew into a wood Denglin（邓林）. And on the spot of his **corpse** rose a huge mountain—Quafu Mountain.

Jingwei Tries to Fill the Sea

The legend, *Jingwei Tries to Fill the Sea* finds its origin in the mythology *Classic of Mountains and Seas*. Once there rested a crow-like bird, Jingwei, with fine feather and red claws, on evergreen mulberry（桑树） in **Fajiu Mountain**. Words said that the bird was the **incarnation** of a girl, Nv Wa（女娃）, the youngest daughter of Emperor Yan, a legendary ruler in ancient Chinese mythology.

Nv Wa loved to appreciate the spectacular nature. One day when she was playing in the east sea, fierce waves **devoured** her. She turned into a bird and mourned herself sadly in the sound "jing-wei, jing-wei". Hence she was named "Jingwei". She hated the sea for taking her life ruthlessly so she decided to fill the

Jingwei Tries to Fill the Sea

sea for a **revenge**. Every day, she flew to and fro between the mountain and the sea, carrying twigs and pebbles and dropping them into the sea. From the myth comes the Chinese idiom "Jingwei tries to fill the sea". It conveys the **dogged** efforts of ancient people to fight against the wild nature. It also indicates their simple idea of cultivating more farming lands by changing the unfavorable **terrains**.

Myth, as a form of folklore, vividly depicts ancient people's struggle against the unpredictable nature and their **tentative** understanding of the universe. It reflects their fear of the unknown, their eagerness to explore nature, and their struggles for survival. Today, some of what they were after are still highly held: longing for health and **longevity**, desire for a harmonious relationship with nature, wish to probe into various mysteries and hope to shape their destiny. Meanwhile, those imaginary stories not only vividly mirrored ancient people's life, but also **nurtured** numerous household literary works, like *Creation of the Gods*[3]. The reading of myth can help people keep well-informed of the past, well-adapted to the present and even well-prepared for the future.

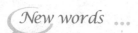

speculation [ˌspekjʊˈleɪʃən] *n.* the act or an instance of speculating 思考；推断

rationalize [ˈræʃənəlaɪz] *v.* If you try to rationalize attitudes or actions that are difficult to accept, you think of reasons to justify or explain them. 使合理化

entity [ˈentɪti] *n.* An entity is something that exists separately from other things and has a clear identity of its own. 实体；实际存在物；本质

punch [pʌntʃ] *v.* If you punch someone or something, you hit them hard with your fist. 用拳猛击

ascend [əˈsend] *v.* If you ascend a hill or staircase, you go up it. 上升；爬坡

murky [ˈmɜː(r)ki] *adj.* A murky place or time of day is dark and rather unpleasant because there is not enough light. 昏暗的；浑浊的

deposit [dɪˈpɒzɪt] *v.* If a substance is deposited somewhere, it is left there as a result of a chemical or geological process. 沉淀

property [ˈprɒpə(r)ti] *n.* The properties of a substance or object are the ways in which it behaves in particular conditions. 特性；属性

antidote [ˈæntɪdoʊt] *n.* An antidote is a chemical substance that stops or controls the effect of a poison. 解药；解毒剂

poisonous [ˈpɔɪzənəs] *adj.* Something that is poisonous will kill you or make you ill if you swallow or absorb it. 有毒的；有害的

refine [rɪˈfaɪn] *v.* If something such as a process, theory, or machine is refined, it is improved by having small changes made to it. 改进；改善

routinize [ruˈtinaɪz] *v.* If you routinize a way of doing something, you make it a normal part of a job or process. 使惯例化；使成常规

deify [ˈdiːɪˌfaɪ] *v.* If someone is deified, they are considered to be a god or are regarded with very great respect. 神话；对…奉若神明

patron [ˈpeɪtrən] *n.* (in ancient Rome) the protector of a dependant or client, often the former master of a freed man still retaining certain rights over him. 保护神

allegedly [əˈledʒɪdlɪ] *adv.* reportedly; supposedly 据说

unrivaled [ʌnˈraɪvəld] *adj.* If you describe something as unrivaled, you are emphasizing that it is better than anything else of the same kind. 无敌的；无双的

scorch [skɔː(r)tʃ] *v.* If something scorches or is scorched, it becomes marked or changes color because it is affected by too much heat or by a chemical. 烤焦；烧焦

indignant [ɪnˈdɪgnənt] *adj.* If you are indignant, you are shocked and angry, because you think that something is unjust or unfair. 义愤填膺的

envelop [ɪnˈveləp] *v.* If one thing envelops another, it covers or surrounds it completely. 包围；笼罩

engulf [ɪnˈgʌlf] *v.* If one thing engulfs another, it completely covers or hides it, often in a sudden and unexpected way. 吞没；吞食

render [ˈrendə(r)] *v.* You can use render with an adjective that describes a particular state to say that someone or something is changed into that state. For example, if someone or something makes a thing harmless, you can say that they render it harmless. 使成为

fatigue [fəˈtiːɡ] *n.* Fatigue is a feeling of extreme physical or mental tiredness. 疲劳

leafy [ˈliːfi] *adj.* Leafy trees and plants have lots of leaves on them. 多叶的；叶茂的

corpse [kɔː(r)ps] *n.* A corpse is a dead body, especially the body of a human being. 尸体

incarnation [ˌɪnkɑː(r)ˈneɪʃən] *n.* An incarnation is an instance of being alive on Earth in a particular from. Some religions believe that people have several incarnations in different forms. 化身

devour [dɪˈvaʊə(r)] *v.* To swallow or eat up greedily or voraciously 吞没

revenge [rɪˈvendʒ] *n.* Revenge involves hurting or punishing someone who has hurt or harmed you. 复仇

dogged [ˈdɒɡɪd] *adj.* If you describe someone's actions as dogged, you mean that they are determined to continue with something even if it becomes difficult or dangerous. 顽强的

terrain [tə(r)ˈreɪn] *n.* Terrain is used to refer to an area of land or a type of land when you are considering its physical features. 地带；地势；地形

tentative [ˈtentətɪv] *adj.* Tentative agreement, plans, or arrangements are not definite or certain, but have been made as a first step. 试探性的；尝试性的

longevity [lɒnˈdʒevɪti] *n.* Longevity is long life. 长寿

nurture [ˈnɜːtʃə(r)] *v.* If you nurture plans, ideas, or people, you encourage them or help them to develop. 培育；滋养

Phrases and expressions

give birth to　产生；引起
be subjected to　受支配；从属于…；遭受
Legend holds that　据传…；据说…
be credited with　认为；把…归功于
to one's memory　纪念
intimidate sb. into doing　恐吓某人做…
talk sb. into doing　说服某人做…
narrow down　缩短
to and fro　来来往往
probe into　探究

Proper names

Classic of Mountains and Seas　《山海经》是记述古代志怪的古籍，战国中后期到汉代初中期的楚国或巴蜀人所作，作者不详。全书现存18篇，内容主要是民间传说中的地理知识。

The Songs of Chu　《楚辞》是屈原创作的一种新诗体，并且也是中国文学史上第一部浪漫主义诗歌总集。

Tale of King Mu of Zhou　《穆天子传》又名《周王传》《穆王传》《周穆王游行记》，是西周的历史典籍之一。

Lvshi Chunqiu　《吕氏春秋》是在秦国丞相吕不韦集合门客们编撰的一部黄老道家名著。

Huainanzi　《淮南子》又名《淮南鸿烈》《刘安子》，是西汉皇族淮南王刘安及其门客集体编写的一部哲学著作，属于杂家作品。

Five Famous Mountains　五岳，指泰山、华山、衡山、嵩山、恒山。

Shennong Herbal Medicine　《神农本草经》又称《本草经》，是中医四大经典著作之一，据说是神农氏所著。

Fajiu Mountain　发鸠山，位于今山西省长治县内。

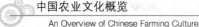

Cultural Notes

1. Acupuncture is a very ancient form of healing. Traditional Chinese Medicine views a person as an energy system in which body and mind are unified, each influencing and balancing the other. This energy flows through the body on channels known as meridians that connect all of our major organs. According to Chinese medical theory, illness arises when the cyclical flow of Qi in the meridians becomes unbalanced or is blocked. Acupuncture is done by inserting sterilized, stainless-steel needles (that are as fine as a human hair) into specific points located near or on the surface of the skin which have the ability to alter various biochemical and physiological conditions in order to treat a wide variety of illnesses. The most significant milestone in the history of Acupuncture occurred during the period of Huang Di—The Yellow Emperor. In a famous dialogue between Huang Di and his physician Qi Bo, they discuss the whole spectrum of the Chinese Medical Arts.

2. Moxibustion (灸) is a traditional Chinese medicine therapy which consists of burning dried mugwort on particular points on the body. It plays an important role in the traditional medical systems of China, Japan, Korea, Vietnam, and Mongolia. Suppliers usually age the mugwort and grind it up to a fluff; practitioners burn the fluff or process it further into a cigar-shaped stick. They can use it indirectly, with acupuncture needles, or burn it on the patient's skin.

3. *Creation of the Gods*, also known by its Chinese names *Fengshen Yanyi* and *Fengshen Bang*, is one of the major vernacular Chinese works in the gods-and-demons (*shenmo*) genre written during Ming dynasty. The story is set in the era of the decline of Shang Dynasty and the rise of Zhou Dynasty. The 100-chapter work intertwines numerous elements of Chinese mythology, including deities, immortals and spirits. The authorship is attributed to Xu Zhonglin.

Chapter 1 Myth and Sacrifice

Exercises

1. Read the following statements and try to decide whether it is true or false according to your understanding.

____ 1) Historically, however, myths constitute a body of speculation that reflects the individual beliefs of a society.

____ 2) Pangu, Chinese Adam called by Westerners, is said to create the world by separating heaven and earth.

____ 3) When Pangu split the egg, light and transparent substances descended while turbid things gradually ascended.

____ 4) It is said that Shennong had a transparent belly, and thus he could taste various herbal medicine.

____ 5) *Shennong Herbal Medicine* is considered to be the earliest medical masterpiece of China.

2. Fill in the blanks with the information you learn from the text.

1) Some of Chinese myths have been passed down even today. We will introduce four legendary figures here: _____, Shennong, _____ and Jingwei.

2) It was said that Shennong invented _____, bent-wood plow, and _____, cut-wood rake, to ease the farming burden and increase working efficiency.

3) It was said that people in Quafu tribe made a name for themselves with their incomparable power and _____.

4) Words said that Jingwei was transformed from the youngest daughter of _____, a legendary ruler in ancient Chinese mythology.

5) _____ indicates the simple idea of ancient people of cultivating farming lands by changing the unfavorable terrains.

3. Explain the following terms.

1) *Pangu Separates Heaven and Earth*
2) *Jingwei Tries to Fill the Sea*

4. Translate the following paragraph into English.

神话是人类社会幼年的产物。它反映了古代劳动人民对自然现象的美妙解释和奇特想象。在他们的心目中，一切自然力量都形象化、人格化了。随后他们又在生产劳动中依照自己的英雄人物形象，创造了很多神的故事，并通过口头流传，这就是神话的起源。阅读神话，不仅可以了解过去，更好地适应当下，甚至还可以做好迎接未来的准备。

5. Critical thinking and discussion.

Myth refers to a popular story accepted as truth or half-truth. It is rich in visual images. And in this sense, myth is largely a kind of product of imagination. How do you define the relationship between imagination-based myth and research-based science? Do you agree that myth can contribute enormously to scientific development? Why or why not?

Section B *Sacrifice*

Sacrifice is the offering of food, objects or animals to gods and spirits as an act of respect or worship. In China it is a blend of ancient religion, **Confucianism**[1], **Taoism**[2] and **Buddhism**[3]. This is because the religious influence in China is far from producing an **overwhelming** mono-religion. As a result, Chinese worship numerous gods and spirits, such as gods of the heaven, the

Ancestor Worship

Chapter 1 Myth and Sacrifice

earth and the ancestors. Sacrifice has **evolved** from a kind of communal behavior to a part of modern civilization and from an unorganized outdoor activity to a regular formal ritual. It gradually developed into a unique tradition featuring both religion and **secularity**. Eventually there appeared a kind of atmosphere, in which a temple stood in every village and every temple burned incense at all seasons.

In ancient times, people believed that there were gods or spirits living somewhere beyond reach who possessed **supernatural** power and used it to reward or punish them. Therefore, they needed to find a way to communicate with gods or spirits about their wishes and requests. So they created sacrifice, through which they could make gods and spirits happy and seek good fortunes from them. Generally, the sacrifice shares some of the following resemblances:

Heaven Worship

- To honor gods with inner respect.
- To behave respectfully: People in the sacrifice would kneel down and koutou in front of the altar to demonstrate their respect and **obedience** to gods or spirits.
- To prepare rich sacrifices: People would present their most valuable things and best stuff they have produced to gods or spirits in order to please gods and spirits, such as grains, silk, wine, vegetables, fruits, jades, blood, pigs, sheep, oxen, living men and women.

- To **chant** to gods or spirits: **Designated** officials would make odes to them, sing highly of their feats, and then speak out the wishes of **mortals**.
- To create certain atmosphere: Candle-burning, incense-burning, tune-playing and dance-making are common elements in sacrifice in order to achieve the desired effect.

Heaven Worship

Heaven worship or heaven sacrifice is a Chinese religious belief earlier than Taoism and Confucianism, later **integrated** with both of them. In Chinese religion, Heaven, or *Tian* （天）, refers to the supreme god of the universe. He is also called **Haotian God**[4] who is believed to have the power to rule the world in some cultural classics. Owing to his **supremacy**, the emperors of China proclaimed themselves to be "Sons of Heaven" in order to maintain and strengthen their **reign** over people. They advocated **Mandate of Heaven**[5] to verify their

Haotian God

legitimacy as rulers and their supposed ability to communicate with Heaven on behalf of their subjects.

Heaven worship appeared side by side with human activities and became highly **ritualistic** in Han Dynasty. Such practice was mostly followed by later dynasties with minor changes. Its process was extremely complex. The emperor would hold official ceremony at altar of heaven （圣坛）, the most famous of which was the **Temple of Heaven**[6] in Beijing. An official institution, the Imperial Board of Astronomy （钦天监）, was to observe the sky, calculate and confirm the exact date for sacrifice. The animals for sacrifice would be carefully chosen and fed 90 days in advance. The emperor and officials would **fast** and wash themselves seven days before the sacrifice. Then the sacrifice would be proceeded roughly as follows:

- To greet Heaven with music （迎神，奏乐）.

- To fire the sacrifices (燔柴). Pigs, sheep, or oxen, or together are set on fire on the altar.
- To salute (行礼). The emperor, crown prince and officials would kneel down **piously** before the altar.
- To present jade and silk to Heaven (奠玉帛). The emperor would wash his hands, ascend the altar, kneel down in front of the **shrine** of Haotian God, offer incense and burn jade and silk.
- The emperor would arrange *zu* (俎), a kind of ritual utensil (进俎).
- The emperor would wash his hands again, clean *jue* (爵), a kind of wine cup, light incense, and make drink offerings (初献).
- The emperor and officials would enjoy the wine and meat offered to Heaven (饮福受胙).
- To remove the sacrifices and utensils (撤豆), such as *dou*.
- To bid farewell to Heaven (送神).
- To burn silk and paper with written prayers at the burning place (去燎所).
- The Emperor would turn around and pay his respect to the burning place halfway to his resting place (望燎).

Zu

Jue

Dou

Heaven or Haotian God, has been the most active, influential and **overawing** god in Chinese ancient life. People would resort to Heaven with any **unresolved** matters, such as marriage, burial, warfare, and the designation of a crown prince. Today, heaven sacrifice is practiced mostly as a way to follow the tradition instead of seeking blessings from him.

Ancestor Worship

Ancestor worship or ancester sacrifice is a ritual practice on the belief that

deceased family members have a continued existence, care the earthly affairs, and possess the ability to influence the living. In Chinese folk religion, man is thought to have multiple souls, *hun* and *po*. Upon death, *hun* and *po* separate. The former ascends into heaven, and the latter **resides** within a spirit tablet （灵牌，灵位）. So the practice of ancestor worship, essentially a family affair, is held in homes and temples with offerings before tablet. The purpose of ancestor worship is to ensure the ancestors' continued well-being and their positive **disposition** towards the living.

Ancestor Sacrifice

Ancestor worship would take place at the funeral of the deceased elders where a home altar is set up by his family members. Sacrifices like food, wine and *Mingbi* （冥币）, would be placed in utensils on the altar or burned in front of the altar. Offerings would be made once or twice a day to ensure the deceased elders to get a good start in the afterlife. Other common customs of Chinese funerals include wearing simple white garb, bowing respectfully, **wailing** dramatically before the altar, having simple meals twice a day and sleeping in a mourning **shack** beside the house. Sometimes, ritual specialists such as Taoist priests or Buddhist monks would be hired to perform specific rituals, often **accompanied** by funeral music performing and scripture chanting to keep evil spirits at bay. The purpose of funerals, regardless of religious beliefs, is to demonstrate respect to the deceased and, provide comfort for them in another world and protect their offspring against evil spirits.

Chapter 1 Myth and Sacrifice

The deceased elders would often be buried with sacrifices, typically things supposed to be needed in the afterlife. This is done as a symbolic demonstration of the **filial piety** of the living to them. Common burial articles include wine, food, as well as candles and incense. For the wealthy and powerful, their corpses would be buried, accompanied by bronze vessels animal or human sacrifices.

Sacrifice was regarded as the top priority in ancient China. The popularity of sacrifice can be fully demonstrated in the emperor's five main responsibilities in reigning a country: managing state affairs, presiding over sacrifices, supporting the people, serving gods and spirits and being present at the homage-paying events from other places. It was held in every season for various purposes in different places. For example, *chunji*（春祭）was practised on the Waking of Insects in spring. On that day the Emperor would plough a certain piece of land in hope of a harvest. Contrastingly, *zhaji*（蜡祭）was held in December. On that day the Emperor would present the yearly gains to thank gods in charge of farming, fishing and hunting. People worship various gods, like the god of wind, rain, thunder and lightening and those who are deified after their death, for example, **Guan Yu**[7] and **Zhong Kui**[8].

Sacrifice appeared and developed with human struggle with nature. It satisfies the dual conceived needs of those gods and spirits, that is, singing their feats and offering them food, drinks and cloths. The main purpose of sacrifice is to ensure the well-being of the living instead of the dead, but sacrifice indeed combines the wishes of the living and the conceived needs of gods and spirits together. Up to now, sacrifice, as a way of age-old social practice, is still practiced in all walks of human life. It can enrich people's knowledge of the physical and mental life of ancient Chinese.

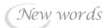

overwhelming [ˌəʊvə(r)ˈwelmɪŋ] *adj.* If something is overwhelming, it affects you very strongly, and you do not know how to deal with it. 势不可挡的；压倒一切的

evolve [ɪˈvɒlv] *v.* If something evolves or you evolve it, it gradually develops over a period of time into something different and usually more advanced. 演变；演化

secularity [ˌsekjəˈlerəti] *n.* the state or quality of being secular 世俗之心；凡俗之心

supernatural [ˌsuːpə(r)ˈnætʃrəl] *adj.* Supernatural creatures, forces, and events are believed by some people to exist or happen, although they are impossible according to scientific laws. 超自然的

obedience [əˈbiːdɪəns] *n.* the state, fact, or an instance of obeying, or a willingness to obey; submission 服从；顺从

chant [tʃɑːnt] *v.* If you chant or if you chant something, you sing a religious song or prayer. 吟颂；咏唱

designate [ˈdezɪˌneɪt] *v.* When you designate someone as something, you formally choose them to do that particular job. 指派；指明

mortal [ˈmɔː(r)tl] *n.* You can describe someone as a mortal when you want to say that he or she is an ordinary person. 凡人

integrate [ˈɪntɪˌɡreɪt] *v.* If you integrate one thing with another, or one thing integrates with another, the two things become closely linked or form part of a whole idea or system. You can also say that two things integrate. 合并；结合

supremacy [suːˈpreməsi] *n.* If one group of people has supremacy over another group, they have more political or military power than the other group. 至高无上

reign [reɪn] *v.* When a king or queen reigns, he or she rules a country. 统治

verify [ˈverɪˌfaɪ] *v.* If you verify something, you state or confirm that it is true. 核实；证明

legitimacy [ləˈdʒɪtəməsi] *n.* the quality or state of being legitimate 合法性；正统

ritualistic [ˌrɪtʃuəˈlɪstɪk] *adj.* Ritualistic acts are the fixed patterns of behavior that form part of a religious service or ceremony. 仪式的；惯例的

Chapter 1 Myth and Sacrifice

fast [fɑːst] *v.* If you fast, you eat no food for a period of time, usually for either religious or medical reasons, or as a protest. 禁食；（尤指）斋戒

pious [ˈpaɪəs] *adj.* having or expressing reverence for a god or gods; religious 虔诚的

shrine [ʃraɪn] *n.* A shrine is a place of worship which is associated with a particular holy person or object. 灵牌；灵位

overaw [ˌəʊvərˈɔː] *v.* If you are overawed by something or someone, you are very impressed by them and a little afraid of them. 吓倒；吓住

unresolved [ˌʌnrɪˈzɒlvd] *adj.* If a problem or difficulty is unresolved, no satisfactory solution has been found to it. 未解决的

deceased [dɪˈsiːst] *adj.* A deceased person is one who has recently died. 过世的；逝去的

reside [rɪˈzaɪd] *v.* If someone resides somewhere, they live there or are staying there. 住；居住

disposition [ˌdɪspəˈzɪʃən] *n.* A disposition to do something is a willingness to do it. 倾向；意向

wail [weɪl] *v.* If someone wails, they make long, loud, high-pitched cries which express sorrow or pain. 痛哭；哀号

shack [ʃæk] *n.* A shack is a simple hut built from tin, wood, or other materials. 窝棚

accompany [əˈkʌmpəni] *v.* If one thing accompanies another, it happens or exists at the same time, or as a result of it. 伴同；给…伴奏

filial [ˈfɪljəl] *adj.* You can use filial to describe the duties, feelings, or relationship which exist between a son or daughter and his or her parents. 子女的

piety [ˈpaɪəti] *n.* Piety is strong religious belief, or behavior that is religious or morally correct. 虔诚；虔敬

Phrases and expressions

be far from 远远不；绝非
sing highly of 赞扬
on behalf of 为了…的利益；代表
on balance 总地来说
resort to 诉诸；采取
preside over 主持；管理
in accordance with 与…一致；依照；秉承

Cultural Notes

1. **Confucianism,** also known as Ruism, is described as a tradition, a philosophy, a religion, a humanistic or rationalistic religion, a way of governing, or simply a way of life. It developed from what was later called the Hundred Schools of Thought from the teachings of the Chinese philosopher Confucius (551-479 BC). It regards the ordinary activities of human life, especially human relationships, as a manifestation of the sacred, because they are the expression of our moral nature, *xing*（性）. Then this worldly concern of Confucianism rests on the belief that human beings are fundamentally good, and teachable, improvable, and perfectible through personal and communal endeavor, especially self-cultivation and self-creation.

2. **Taoism,** also known as Daoism, is a religious or philosophical tradition of Chinese origin which emphasizes living in harmony with the *Tao*. The *Tao* is a fundamental idea in most Chinese philosophical schools. Taoist ethics vary depending on the particular school, but in general tend to emphasize wuwei（无为）: naturalness, simplicity, spontaneity, and the Three Treasures: compassion（慈）, frugality（俭）, and humility（谦）. The roots of Taoism go back at least to the

4th century BC. Early Taoism drew its cosmological notions from the School of Yinyang (阴阳家), and was deeply influenced by one of the oldest texts of Chinese culture, the *Yijing* (《易经》), which expounds a philosophical system about how to keep human behavior in accordance with the alternating cycles of nature.

3. Buddhism is a religion that encompasses a variety of traditions, beliefs and spiritual practices, largely based on original teachings attributed to the Buddha and resulting interpreted philosophies. Buddhism originated in Ancient India sometime between the 6th and 4th centuries BC, from where it spreads through much of Asia.

4. Haotian God is the title of the holy God in Chinese mythology. First documented in Zhou Dynasty, he is equated to the "Heaven" or "Tian", as the God in the sacrifices of different dynasties in China. He is said to be the ruler of the Nature and the earthly kingdoms, together with helpers: the Sun, the Moon, the Wind and the Rain.

5. Mandate of Heaven or Tian Ming is a Chinese political and religious doctrine used to justify the rule of the Emperor of China. According to this belief, heaven embodying the natural order and will of the universe, bestows the mandate on a just ruler of China, the "Heavenly Son" of the "Celestial Empire". The concept was first used to support the rule of the kings of Zhou dynasty (1046-256 BC), and legitimize their overthrow of the earlier Shang dynasty (1600-1069 BC).

6. Temple of Heaven is an imperial complex of religious buildings situated in the southeastern part of central Beijing with an area of 2.7 million square meters. The complex was built in 1420, the 18th year of the reign of Ming Yongle Emperor, and visited by the Emperors of the Ming and Qing Dynasties for annual ceremonies of prayer to Heaven for good harvest.

7. Guan Yu was a general serving under the warlord Liu Bei in the late Eastern Han Dynasty. As one of the best known Chinese historical figures throughout East Asia, Guan Yu's true life stories have largely given way to fictionalized ones, most of which are found in the 14th-century

中国农业文化概览
An Overview of Chinese Farming Culture

historical novel *Romance of the Three King-doms*. He was deified as early as the Sui Dynasty and is still worshipped by many Chinese people today in mainland China, Taiwan, Hong Kong, and among many overseas Chinese communities. In religious devotion he is reverentially called the "Divus Guan" (Guan Di) or "Lord Guan" (Guan Gong).

8. **Zhong Kui** is a figure of Chinese mythology. He is traditionally regarded as a god of driving away the ghosts and evil beings, and reputedly able to command 80,000 demons. His image is often painted on household gates as a guardian spirit, as well as in places of business where high-value goods are involved.

Exercises ...

1. Read the following statements and try to decide whether it is true or false according to your understanding.

____ 1) Chinese worship numerous gods and spirits not including their ancestors.

____ 2) In China, people practice sacrifices in order to ask for something from gods and spirits.

____ 3) In ancient times, people presented their most valuable things to gods and spirits in sacrifices.

____ 4) People would burn some paper money, called *Ming Bi*, in the sacrifice of the deceased relatives.

____ 5) The deceased would often be buried with sacrifices in order to demonstrate the respect from the living.

2. Fill in the blanks with the information you learn from the text.

1) In order to legitimate status as the ruler of a nation, the Emperor of China entitled himself as the "Son of Heaven" based on _____.

2) In Chinese folk religion, man is thought to have multiple souls, *hun*

and *po*. Upon death, the former ascends into _____, and the latter resides within a _____.

3) Sometimes, ritual specialists such as Taoist priests or Buddhist monks would be hired to perform specific rites, often accompanied by funeral music playing and _____ to keep evil spirits at bay.

4) The sacrifice on the Waking of Insects in spring was called _____ and that held in December was called _____.

5) Chinese also worship the god of Wind, Rain, _____ and the ones who are deified after their death, for example, Guan Yu and Zhong Kui.

3. Explain the following terms.

1) Heaven sacrifice
2) Ancestor sacrifice

4. Translate the following paragraph into English.

祭祀是华夏礼典的一部分，也是儒家礼仪的主要组成部分。"祭祀"意为敬神、求神和祭拜祖先。原始时代，人们敬畏大自然的力量，认为有超自然的力量掌控人类的命运。祭祀就是古代人类试图与超自然力量交流的仪式。最初的祭祀活动比较简单，也比较野蛮。人们用竹木或泥土塑造神灵肖像，或在石岩上画出日月星辰、野兽等神灵形象，然后在肖像前陈列献给神灵的食物和其他礼物，对着神灵唱歌、跳舞表示崇敬并求保佑。

5. Critical thinking and discussion.

Filial piety is an important part of traditional Chinese virtues. Some consider it is a kind of manifestation of filial piety to kneel down before the burying places of their decreased, burn incense, paper money, paper cars and houses on Tomb-Sweeping Day every April. While others thinks it is superstition. How do you comment on this phenomenon? And how do you think the culture of filial piety is inherited?

Chapter 2
Wisdom and Thoughts of Farming

Section A Agricultural Policies and Reforms

Agriculture has been a vital sector in China and has been feeding Chinese people since ancient times. In the long history of agricultural practice, the economy pattern of "men **tilling** and women weaving" shaped and developed into agriculture-oriented thought. So the **successive** dynasties have adopted farming-related policies and reforms to promote agricultural development, whether outperformed Shang Dynasty or world-renowned Tang Dynasty. After the foundation of People's Republic of China in modern time, the government has been committed to improving farming productivity by **modifying** agricultural policies.

Well-field System

The Well-field System was a Chinese land distribution method **roughly** from Shang Dynasty to the end of the Warring States Period. Its name came from Chinese character *jing* (井), looking like the sign "#". This character represented the land division: a unit of land was divided into nine equally-sized sections. Its eight outer sections were privately cultivated by eight individual households and the produce was entirely theirs. But the eight outer sections were only allowed to be used but not sold, whose right of use could be passed to the son after the father's death. While the public or center section was communally cultivated by these

Well-field System

eight families on behalf of **aristocrats** and part of the harvest was taken to the country as a tax. It must be said that the private sections could be cultivated only after the public section was finished according to the national law.

The Well-field System was **substantially** a state-owned land system. It effectively avoided land **annexation**, partly helping farmers achieve the dream of "land to the tiller" to some degree. But it gradually fell into disuse in the late Spring and Autumn Period, as the extensive use of ironware and cattle ploughing made it possible for individual household farming to replace collective work. The improvement of farming technology enabled serfs to cut off their links to those aristocrats and practice the independent mode of farming. Therefore, the collapse of Well-field System was **inevitable** as the result of social advance. However, as a system of equal division of land, it still remained ideal to the later generations.

Reforms of Lord Shang Yang

Lord Shang Yang was an outstanding Chinese statesman and thinker in the Warring States Period. With the support of **Duke Xiao of Qin**[1], Shang Yang **implemented** a series of measures to strengthen the power of the state: **abolish** the Well-field System, privatize land, reward the farmers who **overfulfilled** production tasks and deprive them of their properties if not fulfilled, etc. He also **stimulated** the cultivation of virgin lands and wastelands, **prioritizing** agriculture but **restraining** commerce though also **recognizing** the

Lord Shang Yang

achievements of especially successful merchants. His farming policies attracted the immigration of farmers from neighboring states, which provided **massive** laborforce for the sparsely-populated Qin.

Through Shang Yang's reforms, the Qin state established the farming-oriented **feudal** system. His reforms pushed the transformation of social system from slave to feudal society. This complied with the development trend of feudal history and satisfied the interests of new landlord class. His policies laid foundation for the **mightiness** of the Qin state, the first-ever centralized feudal state in Chinese history. But his reforms were opposed by the **privileged** class to some extent, for this damaged their interests.

Equal-field System

The Equal-field System or Land-equalization System was a system of land **allocation** according to populations in China from the Northern Wei Dynasty to mid-Tang Dynasty. It was put into practice around 485 by Emperor Xiaowen of the Northern Wei Dynasty. Afterwards it was adopted by other kingdoms and continued through Sui and Tang Dynasties.

The system worked on the basis of the policy that all land was owned by the government, which would be distributed to individual families. Every individual, including serfs, was given a certain amount of land, depending on his or her working capability. For example, able-bodied men received 40 *mu* (亩, a measurement unit of area in ancient China) while women received 20 *mu*, and each household with farm cattle was granted more land. After land owners passed away, the land would be taken back by the state for redistribution, while the mulberry land could be **inherited** by their offspring.

The system guaranteed farmers' right to use land and reduced the land disputes. It encouraged the effective use of land and ensured that no **arable** land would lay idle. Thus the arable land was expanded. Therefore, the system played a positive role in the recovery and development of agricultural production. Besides, this system prevented aristocrats from controlling more land to expand their power, and allowed common people to make full use of land and secure their livelihood. The above measures increased the government's tax **revenue** and **curbed** aristocrats' land **enclosure**.

Reforms of Wang Anshi

Wang Anshi was a **prominent** economist and statesman of Song Dynasty.

Chapter 2 Wisdom and Thoughts of Farming

In 1067 he attempted a series of major socioeconomic reforms known as the New Policies.

The Law of Farmland and Water Conservancy Engineering（农田水利法）**stipulated** that all households should be engaged in irrigation projects in every part of country by using the materials from local governments. If the materials were not sufficient, they could get a loan from the Zhou（州）or county government; if one Zhou or county failed to accomplish the work alone, it could **unite** neighboring Zhou or counties to complete it. The Law was carried out for seven years, during which more than 1,000 irrigation projects were constructed and 360,000 *qing*（顷, a measurement unit of area in ancient China）farmlands could be irrigated.

Wang Anshi

The Decree of Equal Tax on Square Farmland（方田均税法）was another move of Wang Anshi's series of reform policies. It consisted of two parts: to measure farmlands and divide those into square farmlands of 1,000 steps in four directions respectively, approximately 44 *qing*, and to specify tax on square farmlands based on **terrain** and land **fertility**. Both were started every September by the **designated** staff of a county and completed till the next March. The Decree was abolished in 1085 because of the fierce resistance from local landlords and state officials.

During 14 years of the reforms, the government officials in different places committed serious neglect of duty, and some went so far as to engage in **corruption**. However, the reforms brought some beneficial effects. More than 20 million *qing* farmlands, about half of the total land at that time, were measured and **assessed**. It was the first time in Chinese history to conduct a national survey of farmlands and set a practical example for following dynasties in dealing with farmland-related issues. The state treasure was increased greatly and farmers' burden relieved to some extent.

Single Whip Law

The Single Whip Law or the Single Whip Reform was a **fiscal** law first practiced during the administration of **Zhang Juzheng**[2], Grand Secretary (首辅) in Ming Dynasty. According to the Single Whip Law, thirty or forty different categories of taxes were reduced into a **streamlined** tax structure. Previously the taxes were paid in kind (grain or other products) or in cash (coin), while during Zhang's time all taxes were paid in coin. Thus people had to sell their produce or products and pay a portion of their **monetary** earnings to the government as tax.

Its merits could be generalized into three points. First, it **temporarily** eased class **contradictions** resulting from taxes and labor, which was **conducive** to agricultural development; second, it weakened the enslavement of peasants by the tax system in that period, which made peasants obtain more freedom to participate in agricultural production; more importantly, the monetization of tax revenue promoted the entry of more farm produce into market and this further broke down the natural economy, which created conditions for the development of industry and commerce. Generally speaking, the Law conformed to the development trend of social economy. However, it essentially served the interests of feudal government. Therefore, peasants were still in the **exploited** position.

Zhang Juzheng

Land System of Heavenly Kingdom

The Land System of Heavenly Kingdom was issued in 1853 in Taiping Heavenly Kingdom founded by **Hong xiuquan**[3]. It advocated that land should be shared, so were clothing and money. With the principle of "all farm land must be tilled by all", land was classified into nine types based on its fertility and was equally and reasonably distributed to each household according to populations regardless of gender. Those who were under 15 were given half of what an adult received. With regard to the distribution of produce, the principle of "nobody should keep produce to himself and everything belongs to the state" was adopted, meaning that every household should turn over their produce to the state, only leaving barely enough grain for the family living.

Hong Xiuquan

The land system **embodied** the farmers' strong desire to root out feudal land ownership and to achieve their dream of equal land and produce distribution. However, its goal had never been **materialized** in reality to realize Chinese traditional idea of average wealth, because it failed to arouse the farmers' **initiatives**. The system was merely a **utopian** fantasy.

Principle of People's Livelihood

San-min Doctrine, or Three People's Principles, was a political philosophy developed by Sun Yat-sen to make China a free, prosperous and powerful nation. The three principles were often explained as nationalism, **democracy**, and people's livelihood.

The Principle of People's Livelihood was directly related to land system. It advocated the equalization of landownership, which was mainly directed at the land problem of the city. Its specific implementation was that the state could assess and take over the private land from landlords, and collect taxes on land based on their quoted price of land. Or the state could buy the land at the same price if necessary. It also advocated that every tiller should own land, which aimed to solve the land problem of countryside. It was realized by two measures, one being that the state granted the arable land to tenants and collected some tax, the other being that the state rented uncultivated land to farmers.

Its purpose was to realize the ideal of "land to the tillers" and avoid **monopo-lization** of economy by individual capital. It aimed to realize the state-owned land system and **alleviate** hostility between the rich and the poor. Generally speaking, the principle reflected the main contradictions in China at that time and expressed people's desires for prosperity. However, the principle had its historical limitations, which were mainly **manifested** in the lack of definite and thorough anti-imperialist and anti-feudal ideas.

Household Contract Responsibility System

Household **Contract** Responsibility System was also known as Contract Responsibility System. It was an agriculture production system **initiated** by a group of farmers in a small village in Anhui Province and later extended to other sectors of economy. The system contained two features. First, farmland was still owned by the state. Second, production and management were **entrusted** to individual farming household through long-term contracts. During the contract period, farmers paid taxes to the state and kept all the other produce for themselves. It allowed households to contract land, machinery and other facilities from collective organizations. Households could make operating decisions independently within the limits set by the contract agreement, and could freely dispose of **surplus** over national and collective quotas. The system was warmly accepted by farmers. By the end of 1983, it had **incorporated** more than 90 percent of the country's farming households.

The system not only promoted agricultural development but also turned out to be the breakthrough of rural reforms. Firstly, it greatly inspired farmers' production initiatives, increased agricultural output, and raised rural productivity. Secondly, a large amount of rural labor force was separated from land cultivation and entered village-run factories or town enterprises, which have **evolved** into an important sector in the rural economy. Thirdly, the system changed production mode and farmers' lifestyle in rural areas. It helped **elevate** farmers from self-sufficient minor producers to commodity producers and managers, thus promoting the development of rural market. Finally, with the rapid development of rural economy, farmers' living standards were improved remarkably, with many of them living a relatively well-off life.

Since the implementation of the system, China's agricultural production has been increased at a higher annual rate than before. Total grain output exceeded about 500 billion kilograms in 1996, making China one of the largest grain producers in the world. Output of cotton, cereals, oil, sugar, meat and milk products increased by several times. Per capita consumption of meat, eggs and milk was either close to or above the world average at that time. Deng Xiaoping, a late Chinese leader, once praised it "a great invention of Chinese farmers." The system replaced the People's Commune, and **demonstrated** the value of unified management and aroused the enthusiasm of farmers.

The above accounts show that Chinese governments have attached great importance to the formulation, modification and implementation of agricultural policies. Although the above agricultural policies and reforms have their historical limitations, all of them have undoubtedly stimulated the agricultural development of China.

New words ...

till [tɪl] *v.* When people till land, they prepare the earth and work on it in order to grow crops. 耕地；犁地

successive [sək'sesɪv] *adj.* Successive means happening or existing one after another without a break. 连续的

modify ['mɒdɪfaɪ] *v.* If you modify something, you change it slightly, usually in order to improve it. 修改；变更；改进

roughly ['rʌfli] *adv.* A rough calculation or guess is approximately correct, but not exact. 大概的，粗略的，不精确的

aristocrat ['ærɪstəkræt] *n.* An aristocrat is someone whose family has a high social rank, especially someone who has a title. 贵族

substantially [səb'stænʃəli] *adv.* If you say that something is substantially another thing, you mean that it is mostly the same as that thing in essence. 基本上

annexation [ˈænekˈseɪʃn] *n.* The annexation of something means acquiring something especially territory by occupation. 吞并；占领

inevitable [ɪnˈevɪtəbl] *adj.* If something is inevitable, it is certain to happen and cannot be prevented or avoided. 不可避免的

implement [ˈɪmplɪmənt] *v.* If you implement something such as a plan, you ensure that what has been planned is done. 实施

abolish [əˈbɒlɪʃ] *v.* If someone in authority abolishes a system or practice, they formally put an end to it. 废除，废止（制度、习俗等）

overfulfill [ˌəʊvəfʊlˈfɪl] *v.* If you overfulfill a task, you do more than you are required. 超额完成

stimulate [ˈstɪmjuleɪt] *v.* To stimulate something means to encourage it to begin or develop further. 激励；刺激

prioritize [praɪˈɒrətaɪz] *v.* If you prioritize something, you treat it as more important than other things. 优先考虑；给…优先权

restrain [rɪˈstreɪn] *v.* To restrain something that is growing or increasing means to prevent it from getting too large. 抑制；限制

recognize [ˈrekəgnaɪz] *v.* If someone says that they recognize something, they acknowledge that it exists or that it is true. 认可；承认

massive [ˈmæsɪv] *adj.* Something that is massive is very large in size, quantity, or extent. 巨大的；庞大的

feudal [ˈfjuːdl] *adj.* Feudal means relating to the system or the time of feudalism. 封建体制的；封建时期的

mightiness [ˈmaɪtɪnɪs] *n.* Mightiness means a strong physical strength or power. 强大

privileged [ˈprɪvəlɪdʒd] *adj.* Privileged person has an advantage that most other people do not have, often because of their wealth or high social class. 有特权的

allocation [ˌæləˈkeɪʃn] *n.* The allocation of something is the decision that it should be given to a particular person or used for a particular purpose. 分配；分派

inherit [ɪnˈherɪt] *v.* If you inherit money or property, you receive it from someone who has died. 继承（金钱或财产）

Chapter 2 Wisdom and Thoughts of Farming

arable [ˈærəbl] *adj.* Arable land is land that is used for growing crops such as wheat and barley etc. （土地）适于耕种的

revenue [ˈrevənjuː] *n.* Revenue is money that a company, organization, or government receives from people. 财政收入

curb [kɜːb] *v.* If you curb something, you control it and keep it within limits. 控制

enclosure [ɪnˈkləʊʒə(r)] *n.* An enclosure is an area of land that is surrounded by a wall or fence and that is used for a particular purpose. 圈占地

prominent [ˈprɒmɪnənt] *adj.* Someone who is prominent is important. 杰出的

stipulate [ˈstɪpjuleɪt] *v.* If you stipulate a condition or stipulate that something must be done, you say clearly that it must be done. 规定；明确要求

unite [juˈnaɪt] *v.* If a group of people or things unite or if something unites them, they join together and act as a group. 使团结；使联合

terrain [təˈreɪn] *n.* Terrain is used to refer to an area of land or a type of land when you are considering its physical features. 地形；地带

fertility [fəˈtɪləti] *n.* Land or soil is able to support the growth of a large number of strong healthy plants. 肥力；肥沃

designate [ˈdezɪgneɪt] *v.* When you designate someone as something, you formally choose them to do that particular job. 指派；选派

corruption [kəˈrʌpʃn] *n.* Corruption is illegal behavior by people in positions of authority or power. 腐败；贪污

assess [əˈses] *v.* When you assess the amount of money that something is worth or should be paid, you calculate or estimate it. 评估

fiscal [ˈfɪsk(ə)l] *adj.* Fiscal is used to describe something that relates to government money or public money, especially taxes. 财政的

streamlined [ˈstriːmlaɪnd] *adj.* Something is streamlined means it is simplified. 经过简化以改善效率的

monetary [ˈmʌnɪtri] *adj.* Monetary means relating to money. 货币的

temporarily [tempəˈrerɪli] *adv.* Something that is temporary lasts for only a limited time. 暂时地；短暂地

contradiction [ˌkɒntrəˈdɪkʃn] *n.* If you describe an aspect of a situation as a contradiction, you mean that it is completely different from other aspects. 矛盾

conducive [kənˈdjuːsɪv] *adj.* If one thing is conducive to another thing, it makes the other thing likely to happen. 有益的

exploit [ɪkˈsplɔɪt] *v.* If you say that someone is exploiting you, you think that they are treating you unfairly by using your work or ideas and giving you very little in return. 剥削；压榨

embody [ɪmˈbɒdɪ] *v.* To embody an idea means to be a symbol or expression of that idea. 体现

materialize [məˈtɪərɪəlaɪz] *v.* If a possible or expected event does not materialize, it does not happen. 变为现实；实现

initiative [ɪˈnɪʃətɪv] *n.* If you have initiative, you have the ability to decide what to do next and to do it, without needing other people to tell you what to do. 主动性；积极性

utopian [juːˈtəʊpɪən] *adj.* Utopian is used to describe political or religious philosophies which claim that it is possible to build a new and perfect society in which everyone is happy. 政治或宗教思想空想的，乌托邦式的

democracy [dɪˈmɒkrəsɪ] *n.* Democracy is a system of government in which people choose their rulers by voting for them in elections. 民主；民主制度

monopolize [məˈnɒpəlaɪz] *v.* If you say that someone monopolizes something, you mean that they have a very large share of it and prevent other people from having a share. 独占；垄断

alleviate [əˈliːvɪeɪt] *v.* If you alleviate pain, suffering, or an unpleasant condition, you make it less intense or severe. 减轻；缓解痛苦等

manifest [ˈmænɪfest] *v.* If you manifest a particular quality, feeling, or illness, or if it manifests itself, it becomes visible or obvious. 显示；显露

contract [ˈkɒntrækt] *n.* A contract is a legal agreement, usually between two companies or between an employer and employee, which involves doing work for a stated sum of money. 合同

initiate [ɪˈnɪʃieɪt] *v.* If you initiate something, you start it or cause it to happen. 开始；发起

entrust [ɪnˈtrʌst] *v.* If you entrust something important to someone or entrust them with it, you make them responsible for looking after it or dealing with it. 委托；交付

surplus [ˈsɜːpləs] *n.* If there is a surplus of something, there is more than is needed. 过剩；剩余量

incorporate [ɪnˈkɔːpəreɪt] *v.* If someone or something is incorporated into a large group, system, or area, they become a part of it. 把⋯合并；使并入

evolve [ɪˈvɒlv] *v.* If something evolves or you evolve it, it gradually develops over a period of time into something different and usually more advanced. （使）逐步发展；（使）演化

elevate [ˈelɪveɪt] *v.* If you elevate something to a higher status, you consider it to be better than it really is. 提高

demonstrate [ˈdemənstreɪt] *v.* To demonstrate a fact means to make it clear to people. 证明；说明；表明

Phrases and expressions

be committed to doing 致力于⋯
on behalf of 为了⋯的利益；代表⋯
fall into disuse 废弃不用
cut off 切断
comply with 遵守；服从
lay foundation for 为⋯奠定基础
in kind 以实物偿付
conform to 顺应；符合

with regard to 关于…
root out 彻底根除
be directed at 指向；针对
operating decisions 经营决策
dispose of 解决；将（某物）处理掉
per capita 每人的；人均的
attach great importance to 非常重视

Proper names

Well-field System 井田制
Equal-field System 均田制
Single Whip Law 一条鞭法
Land System of Heavenly Kingdom 天朝田亩制度
Contract Responsibility System 家庭联产承包责任制

Cultural Notes

1. Duke Xiao of Qin（秦孝公，381-338 BC）was the ruler of the Qin state from 361 to 338 BC during the Warring States Period. Duke Xiao was best known for employing the statesman Shang Yang from Wei State, and authorizing him to conduct a series of ground-breaking political, military and economic reforms in Qin. Although the reforms were controversial and drew violent opposition from many Qin politicians, Duke Xiao fully supported Shang Yang and the reforms did help to transform Qin into a dominant superpower among the Seven Warring States.

2. Zhang Juzhen（张居正，1525-1582）was a Chinese reformer and

Chapter 2 Wisdom and Thoughts of Farming

statesman who served as Grand Secretary (首辅) in the late Ming Dynasty during the reigns of the Longqing (隆庆帝) and Wanli (万历帝) emperors.

3. Hong Xiuquan (洪秀全, 1814-1864), born Hong Huoxiu and with the courtesy name Renkun, was a Hakka (客家人) Chinese leader of the Taiping Rebellion against Qing Dynasty, establishing the Taiping Heavenly Kingdom over varying portions of southern China, with himself as the "Heavenly King" and self-proclaimed brother of Jesus Christ.

Exercises

1. Read the following statements and try to decide whether it is true or false according to your understanding.

____ 1) The Well-field System is essentially private ownership of land in Chinese ancient society because it achieved partly the dream of "land to the tiller".

____ 2) The Reform of Lord Shang Yang helped to promote the state of Qin into a powerful state in the late Warring State Period.

____ 3) According to the Equal-field System, farmers could inherit all the land from their ancestors.

____ 4) The new laws put forward by Wang Anshi were supported by the aristocrats of the whole country.

____ 5) The Land System of Heavenly Kingdom was not successfully implemented in the Taiping Peasant Movement.

2. Fill in the blanks with the information you learn from the text.

1) The Reform of Shang Yang was taken into practice during the reign of _____.

2) The Equal-field System was practiced by Emperor Xiaowen in _____.

3) The Law of Farmland and Water Conservancy Engineering by Wang Anshi ordered every household to engage in irrigation works by using the materials received from _____ .

4) The purpose of the Principle of the People's Livelihood is to transform the old private land system to _____ .

5) The first experiment place of Contract Responsibility System was located in _____ county in Anhui province.

3. Explain the following terms.

1) The Well-field System
2) People's Livelihood Principle of Sun Yat-sen

4. Translate the following paragraph into English.

王安石是北宋著名的思想家、政治家、文学家和改革家。宋神宗时期，王安石在全国范围内推行新法，开始大规模的改革运动，涉及政治、经济、军事和社会各方面。王安石变法的目的在于富国强兵，借以改变宋朝极贫极弱的局势。然而，变法由于触动了大地主们的根本利益，遭到了他们的强烈反对，最终失败了。尽管如此，王安石变法仍对当时北宋的社会进步起到了积极作用。

5. Critical thinking and discussion.

Do you think there are differences between the People's Livelihood Principle of Sun Yat-sen and the Contract Responsibility System? In your opinion, which one is more beneficial to people's interest? And why?

Section B Ancient Agricultural Books

Agriculture was the material basis for the existence and development of ancient Chinese civilization. So both government and common people attached great importance to recording and **popularizing** the experience of agricultural production in the past dynasties. Against such cultural background, many agricultural books appeared in ancient China. They collected the farming experience and technology of Chinese working people during thousands of years. According to *Bibliography of Chinese Agriculture*, a total of 500 agricultural books were in Chinese agricultural history, about 300 of which have been passed down so far. They not only embodied the wisdom and thoughts of ancient Chinese but provided references for modern agriculture. Among them, **Book of Fan Shengzhi**, **Qimin Yaoshu**, **Chen Fu Nongshu**, **Wang Zhen Nongshu** and **Nongzheng Quanshu** are the most typical and influential at home and abroad.

Book of Fan Shengzhi

Book of Fan Shengzhi, written by **Fan Shengzhi**[1] during the late Western Han Dynasty, is generally viewed as the oldest existing agricultural **treatise** of China. It devoted to two-thousand-year experience of agricultural production in **Guanzhong Plain** and the middle and lower reaches of the Yellow River. In the book, two basic principles of cultivation were introduced: one was that season and soil mattered in agricultural activities. Under this principle, the irrigation with agricultural **manure** and timely sowing were **highlighted**. The other was early hoeing and early harvesting.

Book of Fan Shengzhi

The book mainly introduced the agricultural production of about 13 crops like millet (稷), wheat, rice, soybean, red bean and so on, specifically including the seed selection, sowing, harvest and seed storage of these crops. It also mentioned the **grafting** of plants (mainly melons) and the methods of land and seed **fertilization**. Besides, the book provided a detailed description of **compartment method** which promised higher yields per acre. It was to plant in square compartments so as to enable plants to better resist **drought** by deep-ploughing. This method was a major innovation of agricultural production technology in Han Dynasty. What's more, the book described the seed coating technique (溲种法), one of the oldest seed processing methods, which referred to covering seed with a layer of fertilizer called seed fertilizer

(种肥). The principle of this method still guides the agricultural production today.

The book illustrated Fan Shengzhi's thoughts of valuing agriculture. In this book, he emphasized the use of scientific production technology in farming. Generally speaking, the book not only summarized the agricultural production experience of **contemporary** northern dryland, but also recorded the southern rice planting methods. It can be inferred that the agricultural production technology of Han Dynasty reached a fairly high level two thousand years ago, which provides the valuable experience for current agricultural production.

Qimin Yaoshu

Qimin Yaoshu, also known as *Important Arts for People's Welfare*, was written by **Jia Sixie**[2] in Northern and Southern Dynasties. He quoted from about 200 ancient books and so **preserved** many fragments of texts that had been otherwise lost, for example, *Book of Fan Shengzhi*.

The book mainly recorded the agricultural production in the lower reaches of Yellow River. It covered a wide range of topics, detailing **agronomy**, **horticulture**, forestry, **sericulture** as well as the processing of **sideline** products, etc. And it also introduced some vegetables and fruits that were rarely cultivated in northern China.

The book introduced many innovative ideas about farming at that time. First, the book stressed the importance for farmers to carefully observe seasons, weather and soil quality, in order that they adjusted their farming activities to these factors. Second,

Qimin Yaoshu

it put emphasis on soil improvement and cultivation techniques by pointing out that proper **moisture** distribution and sufficient fertility in soil were essential. For example, the deep-ploughing methods had been developed for the **amelioration** of soil. Crop **rotation** also helped to improve soil quality, which was also profitable to weed field during different periods of cultivation. Third, the book specified different methods of plant **propagation** like cutting（扦插）, layering（压条）, dividing（分株）and grafting. In particular, the book documented the "**nine brew liquor method**[3]", a brewing method through continuously casting ma-terials, which pioneered the deep **culture** of mildew（霉）and improved alcohol **concentration**. It was of great significance in the history of Chinese alcohol making.

Generally speaking, Jia Sixie **under-scored** the importance of agriculture for the well-being of society in the book. It touched upon broader topics compared with the previous similar works. It included not only the cultivation of plants but cattle breeding and farm produce processing. It recorded some more creative ideas about farming than those of foreign countries at the same period of time, for example, the green manure crop rotation technique was more than 1,200 years earlier than that in Europe. It also offered a great deal of first-hand information about agricultural production and cultivation technology. In addition, it analyzed the ways of increasing farmers' earnings by detailing agricultural production and farming technology. It is acknowledged as the best-preserved agricultural **encyclopedia** still existing in China. It offers a very important source for learning the early agriculture in China.

Chen Fu Nongshu

Chen Fu Nongshu, also *Chen Fu's Book on Agriculture*, was **compiled** by Chen Fu[4] and finished in 1149. It covered almost all aspects of agriculture, including three volumes: rice cultivation technology, water buffalo raising and sericulture, with the first volume as the core. In the book the rice fields in the south were divided into four types, namely, early-cropping rice fields, late-cropping rice fields, cold water fields in mountain, and lowland fields. And the technologies about plowing, draining and sunning of rice fields were detailed.

The book pointed out a lot of significant thoughts about agriculture as follows:

- The observation of weather and seasons was the basic duty of all farmers. Farmers had to follow natural laws rather than rely on pure chance;
- Infertile soil, or even **exhausted** fields, could regain their productivity with fertilizers. This statement **refuted** the old proverb that three or five years of cultivation would result in the infertility of the soil;
- Farmers should only invest in fields where they would **accordingly** gain a rich harvest;

Chen Fu Nongshu

- Farmers should plan the cultivation of crops in advance in order to obtain the highest yield and prepare for **combating** unfavorable conditions.

Chen Fu Nongshu had the following contributions to China's agricultural production. First, it was the earliest book on the agricultural experiences of southerners. While the previous ones were mostly about that of the northerners in the Yellow River basin. Second, it recorded a lot of practical and reliable information from Chen Fu's own experience as a farmer, such as the amelioration of soil, the use of manure, and the raising of rice shoots. Third, it embodied the style of modern **ecological** agriculture, for example, mulberry trees could be planted on the levee of the pond, fish raised in the pond, and water used for **irrigation**, which made agriculture, fishery and sideline develop at the same time. What's more, it discussed many topics about agriculture for the first time, such as the technology of rice seedling nursery and the agricultural management in agricultural production.

Wang Zhen Nongshu

Wang Zhen Nongshu, also *Wang Zhen's Book on Agriculture*, was written by **Wang Zhen**[5] and was finished in 1313. It included three parts: universal principles for agriculture and sericulture, crops and farm tools. In the first part, it discussed the following topics: cattle ploughing, sericulture, fruit cultivation, poultry raising, the production techniques of field crops and so on. In the second part, it informed about the types of crops and cultivation methods in different regions. More importantly, it included crops from Southeast Asia and other foreign countries like fragrant rice (香稻), watermelons and spinach which were introduced to China during Yuan Dynasty. The third part listed more than 300 farm tools **illustrated** with texts about the origin and the use of various tools in agriculture.

Wang Zhen Nongshu

This book shared an important position in the history of agricultural science and technology in China. To begin with, it featured the knowledge of ancient tools, which was rarely mentioned in previous agricultural books. This marked that Chinese traditional farm tools had been developed to a high level. The illustrations of the tools made the book a real treasure of agriculture in ancient China. More importantly, it not only covered the information on agricultural techniques in both northern and southern China but their similarities and differences for the first time, pointing out that both should learn from each other to promote the development of agriculture. It was also the first book that described almost all methods of soil use, such as polders (圩田), floating farmland (架田), terrace fields (梯田), muddy fields (涂田), and the like. So to this day it remains a significant source for the ancient agriculture in northern and southern China.

Nongzheng Quanshu

Nongzheng Quanshu, also *Complete Treatise on Agriculture*, was written by **Xu Guangqi**[6]. It was not actually completed during his lifetime and left to his disciple Chen Zilong (陈子龙) who **amended** the text and recompiled it. It completely reviewed all aspects of farming in ancient China and provided evidence for the agricultural level of China in the 17th century.

The book discussed land **reclamation**, water conservancy and **famine** relief. It mentioned Xu Guangqi's thought that people should reclaim wasteland and establish water conservancy projects. By doing so they could cultivate grain instead of purchasing it from southeast, which could turn northwest into a major grain yielding area. Particularly, the book **foregrounded** the issue of the famine relief as an independent part. It reviewed the famine policies of past dynasties, statistically analyzed droughts, plagues, as well as their relief measures, and finally suggested a lot of **edible** plants and vegetables for famine. It recorded Xu Guanqi's study of the **lean** years in Chinese history and even the concrete situations of locust（蝗虫）plagues which had ever happened in China. According to the book, the most effective way of putting an end to lean years was to construct water conservancy projects on a large scale.

Nongzheng Quanshu

The book was different from other agricultural books in two aspects. First, the book featured Xu Guangqi's ideology of agricultural administration which advocated that agriculture development was the root of a wealthy country. This was the basic idea that ran through the book. For example, the measure of delaying grain taxes was enforced to encourage **refugees** back home to cultivate land. While other books, whether it was *Qimin Yaoshu* or *Wangzhen Nongshu*, centered on the agriculture-based thoughts with the emphasis on the production technology and knowledge. These books could be said to be purely technical agricultural books. Second, the book quoted a large amount of previous literatures combined with Xu Guangqi's own practical experience and research in agriculture, for example, the extensive irrigation ideas from *Wangzhen Nongshu*. It was a great agricultural work of special importance in Ming Dynasty. And even in modern China, this book is highly valued as a significant agricultural encyclopedia.

The above five books are masterpieces of ancient agricultural treatises in China. And they reflect the wisdom and thoughts of working people in their long-term agricultural practice. More important, they also include important agricultural ideas of the writers. They show that ancient Chinese agriculture reached **brilliant** achievements, as remarkable as traditional Chinese medicine, **astronomy** and mathematics. As is known to all, history is a great invisible wealth. So these precious books can serve as a mirror for modern agricultural development and may offer some solutions to some problems in its development.

New words

popularize ['pɒpjələraɪz] *v.* To popularize an academic subject or scientific idea means to make it easier to understand for ordinary people. 普及；推广

treatise ['triːtɪs] *n.* A treatise is a long, formal piece of writing about a particular subject. 专题论文

manure [mə'njʊə] *n.* Manure is animal faeces, sometimes mixed with chemicals, that is spread on the ground in order to make plants grow healthy and strong. 粪肥；肥料

highlight ['haɪlaɪt] *v.* If someone or something highlights a point or problem, they emphasize it or make you think about it. 使突出；强调；使注意

graft [grɑːft] *v.* If a part of one plant or tree is grafted onto another plant or tree, they are joined together so that they will become one plant or tree, often in order to produce a new variety. 将…嫁接到

fertilization [ˌfɜːtəlaɪ'zeɪʃn] *n.* Fertilization means to improve soil quality in order to make plants grow well on it, by spreading solid animal waste or a chemical mixture on it. 施肥

drought [draʊt] *n.* A drought is a long period of time during which no rain falls. 久旱；旱灾

contemporary [kən'temprəri] *adj.* Contemporary people or things happened at the same time as something else you are talking about. 同时期的；同时代的

An Overview of Chinese Farming Culture

preserve [prɪ'zɜːv] *v.* If you preserve something, you take action to save it or protect it from damage or decay. 保护；保存

agronomy [ə'grɒnəmɪ] *n.* Agronomy is the study of the application of soil and plant sciences to land management and crop production. 农学；农艺学

horticulture ['hɔːtɪkʌltʃə] *n.* Horticulture is the study and practice of growing plants. 园艺学

sericulture ['serɪˌkʌltʃə] *n.* Sericulture means raising silkworms in order to obtain raw silk. 养蚕；蚕事

sideline ['saɪdlaɪn] *n.* A sideline is something that you do in addition to your main job in order to earn extra money. 副业

moisture ['mɔɪstʃə(r)] *n.* Moisture is tiny drops of water in the air, on a surface, or in the ground. 水分

amelioration [əˌmiːlɪəˈreɪʃn] *n.* The amelioration of a situation means it made or better or easier. 改善；使变好

rotation [rəʊ'teɪʃn] *n.* The rotation of a group of things or people is the fact of them taking turns to do a particular job or serve a particular purpose. 轮流

propagation [ˌprɒpə'geɪʃn] *n.* The propagation of plants means more of them are grown from the original ones. 繁殖

culture ['kʌltʃə(r)] *n.* In science, the culture of a group of bacteria or cells means to grow them. 培养；培植

concentration [ˌkɒnsn'treɪʃn] *n.* The concentration of a substance is the proportion of essential ingredients or substances in it. 浓度

underscore [ˌʌndə'skɔː(r)] *v.* If something such as an action or an event underscores another, it draws attention to the other thing and emphasizes its importance. 强调；突出

encyclopedia [ɪnˌsaɪklə'piːdɪə] *n.* An encyclopedia is a book or set of books in which facts about many different subjects or about one particular subject are arranged for reference. 百科全书

compile [kəm'paɪl] *v.* When you compile something such as a report, book, or program, you produce it by collecting and putting together many pieces of information. 汇编；编纂

exhausted [ɪɡˈzɔːstɪd] *adj.* Being exhausted means being completely used or finished. 耗尽的；枯竭的

refute [rɪˈfjuːt] *v.* If you refute an argument, accusation, or theory, you prove that it is wrong or untrue.

驳倒；驳斥

accordingly [əˈkɔːdɪŋli] *adv.* You use accordingly to introduce a fact or situation which is a result of something that you have just referred to. 因此；于是

combat [ˈkɒmbæt] *v.* If people in authority combat something, they try to stop it happening. 斗争；打击

ecological [ˌiːkəˈlɒdʒɪkl] *adj.* Ecological means involved with or concerning ecology. 生态的

irrigation [ˌɪrɪˈɡeɪʃn] *n.* Irrigation means supplying dry land with water by means of ditches etc. 灌溉

illustrated [ˈɪləstreɪtɪd] *adj.* If a book or something is illustrated, some pictures, photographs or diagrams are put into it. 加了插图或图表的

amend [əˈmend] *v.* If you amend something that has been written such as a law, or something that is said, you change it in order to improve it or make it more accurate. 修改；修订

reclamation [ˌrekləˈmeɪʃn] *n.* Reclamation is the process of changing land that is unsuitable for farming or building into land that can be used. （荒地等）开垦；改造

famine [ˈfæmɪn] *n.* Famine is a situation in which large numbers of people have little or no food, and many of them die. 饥荒

foreground [ˈfɔːɡraʊnd] *v.* To foreground certain features of a situation means to make them the most important part of a description or account. 强调；突出

edible [ˈedəbl] *adj.* If something is edible, it is safe to eat and not poisonous. 可食用的；能吃的

lean [liːn] *adj.* A lean time mean that people have less of something such as money. 萧条的；不景气的

refugee [ˌrefjuˈdʒiː] *n.* Refugees are people who have been forced to leave their homes or their country. 难民；寻求庇护者

brilliant [ˈbrɪliənt] *adj.* You can say that something is brilliant when you are very pleased about it or think that it is very good. 很棒的；出色的

astronomy [əˈstrɒnəmi] *n.* Astronomy is the scientific study of the stars, planets, and other natural objects in space. 天文学

Phrases and expressions

pass down 把…一代传一代，使流传
devote to 完全用于（某事或做某事）；致力于…
put emphasis on 强调；重视
be of great significance 非常重要的；具有重大意义的
touch upon 论及；简单涉及
be acknowledged as 被认为…；被承认为…
pure chance 纯属偶然的机会
on a large scale 大规模地
run through 贯穿

Proper names

Book of Fan Shengzhi 《氾胜之书》
Qimin Yaoshu 《齐民要术》
Chen Fu Nongshu 《陈敷农书》
Wang Zhen Nongshu 《王祯农书》

Chapter 2　Wisdom and Thoughts of Farming

Nongzheng Quanshu　《农政全书》

Guanzhong Plain　关中平原，又称渭河平原（Weihe Plain），位于陕西省中部，介于秦岭和渭北北山之间。

compartment method　区田法

Cultural Notes ...

　　1. Fan Shengzhi（氾胜之）was also named Fan Sheng, who was born in modern Caoxian County（曹县）of Shandong province. He was a famous ancient agronomist about at the end of Western Han period. He guided the agricultural production in modern Shaanxi Province（陕西）as an officer and so he knew farming and had a lot of experiences in agricultural production.

　　2. Jia Sixie（贾思勰）was born in modern Shouguang County（寿光）of Shandong province. He was a famous ancient agronomist. He went to many places such as Shandong Province, Hebei Province and Henan Province after he was a court official. In these places, he carefully observed and studied the local agriculture production technology and consulted some experienced farmers so as to obtain a great deal of agricultural production knowledge. This established the foundation for the writing of *Qimin Yaoshu*, which was finished between about Beiwei Dynasty and Dongwei Dynasty.

　　3. Nine brew liquor method（九酝酒法）was the oldest Chinese liquor brewing method with written records according to *Qimin Yaoshu*. The book recorded that the method originated from the brewing liquor technology of Bozhou（亳州）in Eastern Han Dynasty. And it was introduced by Cao Cao（曹操）when he presented "nine brew liquor" produced in spring to Emperor Xian of Han（汉献帝）. Cao Cao found that the liquor would be bitter if nine *hu*（斛, a name of measuring vessel of ancient China）of rice was used, so ten *hu* was used later, then the liquor became perfect. This method was inherited by the current Gujinggong liquor (古井贡酒) of Anhui province.

4. Chen Fu (陈敷, 1076-?), was a hermit and agronomist in Song Dynasty. He was engaged in agriculture as a farmer in modern Yizheng County (仪征) of Jiangsu Province and finished the writing of *Nongshu* at the age of 74.

5. Wang Zhen (王祯, 1271-1368) was born in modern Dongping County (东平) of Shandong Province, an agriculturalist and agricultural mechanist in Yuan Dynasty. When he was a district magistrate he actively promoted the amelioration of agricultural tools, raising methods for trees, hemp and cotton, which provides the rich experiences for the writing of the agriculture book.

6. Xu Guangqi (徐光启, 1562-1633) was born in modern Shanghai city, a famous scientist and statesman in Ming period. He was devoted to the research of mathematics, astronomy, calendar, water conservancy and other aspects especially in the field of agriculture. Meantime, he was a forerunner to communicate with the western cultures which made an important contribution to the cultural exchange between China and the west in 17th century.

Exercises

1. Read the following statements and try to decide whether it is true or false according to your understanding.

____ 1) It is generally agreed that the *Book of Fan Shengzhi* is the oldest agricultural book in China.

____ 2) *Qimin Yaoshu* mainly recorded the agricultural production in the Changjiang River basin.

____ 3) *Chen Fu Nongshu* is the earliest agricultural book to touch upon the southern Chinese agriculture with the rice production technology as the core.

____ 4) *Wang Zhen Nongshu* mainly centered on the farming not in northern

China but in southern China.

____ 5) It is in the agricultural book *Nongzheng Quanshu* that Xu Guangqi expressed his thought that agricultural administration is very important to run a country.

2. Fill in the blanks with the information you learn from the text.

1) The *Book of Fan Shengzhi* provided an exact description of the deep-ploughing method, by which the soil was enabled to preserve moisture and fertility, called _____ .

2) The "nine brew liquor method", which pioneered the deep culture of mildew and improved the alcohol concentration, was of great significance in the history of Chinese _____ .

3) The agricultural book is _____ which was dedicated to the water buffalo, its use and medical care.

4) *Wang Zhen Nongshu* mainly narrated the universal principles for agriculture and sericulture, the theory of crops and _____ .

5) Like previous agricultural books, *Nongzheng Quanshu* discussed the agricultural production and technology, particularly _____ which was rare in other similar-kind books.

3. Explain the following terms.

1) *Qimin Yaoshu*

2) *Nongzheng Quanshu*

4. Translate the following paragraph into English.

贾思勰是我国古代一位著名的农学家，他强调农业生产遵循自然规律，农作物必须因地种植，不误农时，还要改革生产技术和工具。他著了一部农书，就是《齐民要术》。这本农书总结了我国北方劳动人民积累的生产经验，主要介绍了农、林、牧、副、渔业的生产方法，被称为我国现存最完整的农业百科全书，是研究我国古代农业生产的一个重要资源。

5. Critical thinking and discussion.

Xu Guangqi put forward his ideology of agricultural administration that agricultural development was the root of a wealthy country in *Nongzheng*

Quanshu. China government proposes to comprehensively promote rural revitalization in 2023. While nowadays the policy of the reforestation of the cultivated land has been put forward in the west of China. Do you think this policy means that China government doesn't attach great importance on the agricultural development? What is your comment on this policy?

Chapter 3
Twenty-Four Solar Terms

Section A General Introduction to Twenty-Four Solar Terms

Ancient Chinese created a traditional calendar, dividing a year into 24 periods, to guide farming activities over the four seasons based on their daily observation of the Sun's **annual** movement. The 24 periods were called 24 *jieqi* (节气) or 24 solar terms. Each solar term has a specific name. These 24 solar terms are Beginning of Spring (*Li Chun*), Rain Water (*Yu Shui*), Insects Awakening (*Jing Zhe*), Spring Equinox (*Chun Fen*), Clear and Bright (*Qing Ming*), Grain Rain (*Gu Yu*), Beginning of Summer (*Li Xia*), Grain Buds (*Xiao Man*), Grain in Ear (*Mang Zhong*), Summer Solstice (*Xia Zhi*), Lesser Heat (*Xiao Shu*), Greater Heat (*Da Shu*), Beginning of Autumn (*Li Qiu*), End of Heat (*Chu Shu*), White Dew (*Bai Lu*), Autumn Equinox (*Qiu Fen*), Cold Dew (*Han Lu*), First Frost (*Shuang Jiang*), Beginning of Winter (*Li Dong*), Light Snow (*Xiao Xue*), Heavy Snow (*Da Xue*), Winter Solstice (*Dong Zhi*), Lesser Cold (*Xiao Han*) and Greater Cold (*Da Han*).

The written records of the 24 solar terms can be dated back to Pre-Qin period. The term *jieqi* was found in Chinese medical classic **Huangdi Neijing**[1]. According to this book, every five days was called a *hou* (候) or five-day parts, and every three *hou* made a *qi* (气). Therefore, a year was divided into 72 *hou*, or 24 *qi*, popularly known as 24 *jieqi*. The Chinese classic **Lvshi Chunqiu**[2] for the first time mentioned the names of 8 solar terms: Beginning of Spring, Beginning of Summer, Beginning of Autumn, Beginning of Winter, Spring Equinox, Summer Solstice, Autumn Equinox and Winter Solstice. These 8 solar terms are the most important ones in the 24 solar terms. The Chinese philosophical book **Huainanzi**[3] in the early Han

Dynasty listed 24 terms identical to those of today. These 24 solar terms were soon **absorbed** into the first official Chinese calendar *Taichu Calendar*[4]. In this calendar, the 24 solar terms began at Winter Solstice. While in the early Qing Dynasty, Spring Equinox was adopted as the first solar term in *Shixian Calendar*[5], which was improved on the basis of *Chongzhen Calendar*[6] with the use of new western astronomical instruments. However, currently Beginning of Spring is **customarily** regarded as the first one because it happens to be the first solar term in spring which signals the beginning of farming activities in a year.

According to **transitions** between four seasons as well as changes in temperature, **precipitation,** and other natural phenomena during a year, the system of the 24 solar terms is **categorized** into 4 groups. The first group includes eight solar terms: Beginning of Spring, Beginning of Summer, Beginning of Autumn, Beginning of Winter, Spring Equinox, Autumn Equinox, Summer Solstice and Winter Solstice. The first four solar terms in this group mark the beginning of four seasons, and the other four not only reflect seasonal changes, but show the changes of height of the Sun. Spring Equinox and Autumn Equinox come when the Sun shines directly at the equator, then day and night are almost equally divided. On the day of Summer Solstice, the Sun travels to its northernmost point, therefore the northern **hemisphere** experiences the longest day, known as *ribeizhi*（日北至）or Sun North Most. While on the day of Winter Solstice the Sun reaches its southernmost point, the northern hemisphere has the longest night, known as *rinanzhi*（日南至）or Sun South Most.

Five solar terms in the second group show the changes of temperature in different periods. They are Lesser Heat, Greater Heat, End of Heat, Lesser Cold, and Greater Cold. Lesser Heat marks the coming of the hottest period. Greater Heat represents the hottest days in a year. End of Heat indicates that the burning hot is fading away and there will be no summer heat since this day. *Sanfu*（三伏）, three periods of the hot season, occurs between Lesser Heart and End of Heat. *San* refers to the Chinese figure "three", while *fu* implies that people had better lie down rather than move in order to hide from the heat. Historically, Chinese people also consider that on the day of Winter Solstice, *shujiu*[7]（数九）begins. After Winter Solstice, there are nine

nine-day periods which mark the changes of temperatures from winter to early spring. Lesser Cold is regarded as the most frozen time of winter and it often falls on the day of the third nine-days. Greater Cold signals the end of the 24 solar terms and the end of winter. Although modern meteorological observation shows that the weather during Greater Cold is not colder than Lesser Cold in some regions of China, the lowest temperatures of the whole year still occur in Greater Cold period in some coastal areas.

The third group includes Rain Water, Grain Rain, White Dew, Cold Dew, First Frost, Light Snow and Heavy Snow. These seven solar terms reflect the changes of precipitation. During Rain Water, the form of precipitation **transforms** from snowfall to rainfall and the amount of rainfall obviously increases. Grain Rain **originates**

from the old saying "rain brings up the growth of hundreds of grains" （雨生百谷）. In the days of Grain Rain, the temperature keeps **elevating** quickly and it rains frequently. During White Dew, the temperature starts to drop rapidly and rain days come. At the very day of White Dew, vapors in the air often **condense** into white dew at night due to big temperature difference of day and night. Cold Dew is a period in which the weather turns from cooling to cold. Rainfall becomes less while the dew is heavier than that of White Dew. First Frost, the last solar term in autumn, means that the vapors in the air freeze and frost begins to appear. Light Snow signals the arrival of snow but only in small quantities. With temperature dropping significantly, Heavy Snow arrives with heavier snowfall. At this time of year, snow is expected in most places of China.

Insects Awakening, Clear and Bright, Grain Buds and Grain in Ear in the last group **embody** natural phenomena. In Chinese folklore, during the period of Insects Awakening, insects **hibernating** in winter are waken up by spring thunder and the earth begins to come back to life. "Clear" and "Bright" describe the clear air and bright scene in spring during this period. From that day on, temperature continues to rise and rainfall starts to increase. Grain Buds and Grain in Ear show the ripeness of crops. Grain Buds represents a period when grains are not totally full and ripe. While Grain in Ear signifies the ripening of crops.

The system of the 24 solar terms originates in the reaches of the Yellow River, which was the center of ancient China's major political, economic, cultural and agricultural activities for a long time. Therefore the 24 solar terms mainly reflect the temperature, precipitation, as well as other natural phenomena in this region. Although it cannot precisely reflect the climate changes in other regions across China, the system is still generally accepted by people as a **timeframe** in their agricultural production as well as daily routines all over the country. Some solar terms like Clear and Bright as well as Winter Solstice have been so influential that Qingming Festival and Winter Solstice Festival evolve from them **respectively** and are highly valued today.

Nowadays the 24 solar terms still play multi-functions in different areas of life. They remain particularly important to farmers in guiding their agricultural practices. Besides the guiding role in people's farming life and daily routines, the 24 solar terms are common themes in literary works, and they are often referred to in nursery rhymes, proverbs as well as poems. **A song for the 24 solar terms**[8] from long time ago is still popular and even taught to children at school. Not only widely accepted and respected in China, the 24 solar terms are recognized and appreciated as well in international community. On November 30, 2016, the 24 solar terms were **inscribed** on the **Representative List of the Intangible Cultural Heritage of Humanity**[9] by the **UNESCO**. This **legacy** reflects Chinese people's respect for tradition and their wisdom to live in harmony with nature. It is also a great contribution to the world's cultural **diversity**. Therefore, it is a precious **heritage** to all the human beings.

New words

annual [ˈænjuəl] *adj.* Annual events happen once every year. 每年一次的

absorb [əbˈzɔːb] *v.* If a group is absorbed into a larger group, it becomes part of the larger group. 使并入

customarily [ˈkʌstəmərəli] *adv.* 通常；习惯上

transition [trænˈzɪʃn] *n.* Transition is the process in which something changes from one state to another. 转型

precipitation [prɪˌsɪpɪˈteɪʃən] *n.* Precipitation is rain, snow, or hail. 降水

categorize [ˈkætəɡəraɪz] *v.* If you categorize people or things, you divide them into sets. 把…分类

hemisphere [ˈhemǝsfɪǝ] *n.* One half of the earth. 半球

transform [trænsˈfɔːm] *v.* To transform something into something else means to change or convert it into that thing. 使改变；使转换

originate [əˈrɪdʒəneɪt] *v.* When something originates, it begins to happen or exist. 起源

elevate [ˈelɪveɪt] *v.* To elevate something means to increase it in amount or intensity. 提高

condense [kənˈdens] *v.* When a gas or vapor condenses, or is condensed, it changes into a liquid. 冷凝

embody [ɪmˈbɒdi] *v.* To embody an idea or quality means to be a symbol or expression of that idea or quality. 体现；具体象征

hibernate [ˈhaɪbəneɪt] *v.* Animals that hibernate spend the winter in a state like a deep sleep. 冬眠

timeframe [ˈtaɪmˈfreɪm] *n.* 时间表

respectively [rɪˈspektɪvli] *adv.* Respectively means in the same order as the items that you have just mentioned. 分别地

inscribe [ɪnˈskraɪb] *v.* If you inscribe words on an object, you write or carve the words on the object. 刻

legacy [ˈleɡəsi] *n.* A legacy is money or property which someone leaves to you when they die. 遗产

中国农业文化概览
An Overview of Chinese Farming Culture

diversity [daɪˈvɜːsɪtɪ] *n.* The diversity of something is the fact that it contains many very different elements. 多样性

heritage [ˈherɪtɪdʒ] *n.* A country's heritage is all the traditions, or features of life there that have continued over many years and have been passed on from one generation to another. 遗产；传统

Phrases and expressions

identical to 同样的；和…相同
fade away 逐渐消失
regard…as 把…认作
refer to 参考
in harmony with 和谐；一致

Proper names

UNESCO Shortened form for United Nations Educational, Scientific, and Cultural Organization. 联合国教科文组织

Cultural Notes

1. *Huangdi Neijing* (《黄帝内经》) was compiled roughly two thousand years ago. This great masterpiece forms the theoretical basis of Traditional Chinese Medicine. Covering not only medicine but also philosophy, sociology, anthropology, military strategy, mathematics, astronomy, meteorology, ecology, this book demonstrates that even in

ancient times, people accomplished scientific achievements that are applicable, relevant, and innovative even in modern times.

2. Lvshi Chunqiu (《吕氏春秋》) is an encyclopedic Chinese classic book compiled around 239 BC under the patronage of prime minister Lv Buwei (吕不韦) of Qin Dynasty. It is also one of the longest of the early texts, consisting of 8 chapters of examinations (*lan* 览), 6 chapters of discourses (*lun* 论), and 16 chapters of almanacs (*ji* 纪). Although it has been compiled by the hands of many authors, the whole composition is very consistent and integrative. It covers a vast range of topics, beginning with the seasons, the corresponding phenology (物候学) and the integrative correlation of all appearances in the universe.

3. Huainanzi (《淮南子》) is a Chinese classic essay collection of various philosophical treatises in the 2nd century BC compiled under the mentorship of Liu An (刘安), Prince of Huainan of Western Han Dynasty. The core concept of *Huainanzi* is that there is a *Tao* (道) in the universe, and many parts of it are influenced by the Taoist book *Zhuangzi* (《庄子》). The chapter *Qisuxun* (齐俗训), for instance, inherits the thoughts of the chapter *Qiwulun* (齐物论) in *Zhuangzi*.

4. Taichu Calendar (太初历) is the first relatively complete calendar in China. According to it, a year had 365.2502 days and a month had 29.53086 days. The calendar adopted the 24 solar terms, which were useful for the farming season.

5. Shixian Calendar (时宪历), also known as the *Chongzhen* (崇祯) *Calendar*, was dedicated to Emperor Chongzhen of Ming Dynasty. It was spread by Emperor Shunzhi (顺治) of Qing dynasty who changed its name to *Shixian Calendar*. The calendar was established by the Jesuit Johann Adam Schall Von Bell (汤若望). It is the last luni-solar calendar in China.

6. Chongzhen Calendar (崇祯历) was compiled by some missionaries in cooperation with Chinese scientists under the reign of Chongzhen Emperor. The compiling of *Chongzhen Calendar* lasted for five years and was completed in 1634. An important part of the whole calendar is the translation of western astronomy theory. *Chongzhen Calendar* was different from previous traditional Chinese calendar in that a whole set of measurement and calculation system different from the traditional astronomy was introduced.

7. *Shujiu*（数九）is nine periods of nine days each after winter solstice, the coldest time of the year. Ancient Chinese wrote a *Shujiu* song. The song says "in the first and second nine days, people keep their hands in pockets; in the third and fourth nine days, people can walk on ice; in the fifth and sixth nine days, people can see willows along the river bank; in the seventh and eighth nine days, the swallow comes back and in the ninth nine days, the yak starts working."（一九二九不出手；三九四九冰上走；五九六九沿河看柳；七九河开八九雁来；九九加一九，耕牛遍地走。）This song reflects changes in temperature and phenology（物候）during this period.

8. **A song for the 24 solar terms**（二十四节气歌）was written by ancient Chinese and the song goes like this：

春雨惊春清谷天，夏满芒夏暑相连。

秋处露秋寒霜降，冬雪雪冬小大寒。

9. **Representative List of the Intangible Cultural Heritage of Humanity**（人类非物质文化遗产代表作名录）was established by the United Nations Educational, Scientific, and Cultural Organization with the aim of ensuring better protection of important intangible cultural heritages worldwide and the awareness of their significance. It contains intangible cultural heritage elements that help demonstrate the diversity of cultural heritage and raise awareness about its importance.

Exercises

1. Read the following statements and try to decide whether it is true or false according to your understanding.

____ 1) Ancient Chinese created a traditional calendar dividing the year into 24 parts based on observation of the sun's annual motion.

_____ 2) The 24 solar terms were inscribed on the Representative List of the Intangible Cultural Heritage of Humanity by UNESCO in 2016.

_____ 3) China's 24 solar terms originated in the region around the Yellow River reaches.

_____ 4) Every five days was called a *hou*, and every three *hou* made a *qi* in the book *Huainanzi*.

_____ 5) Spring Equinox was adopted as the first solar term since the 24 solar terms were created.

2. Fill in the blanks with the information you learn from the text.

1) On the day of Winter Solstice the sun reaches its southernmost point, the _____ has the longest night, known as *rinanzhi* by Chinese ancients.

2) In Chinese folklore, during the period of _____, animals hibernating in winter wake up by spring thunder and that the earth begins to come back to life.

3) Grain Buds and _____ show the ripeness of crops.

4) _____ and Clear and Bright are both Chinese traditional festivals and significant solar terms.

5) The Chinese philosophical book _____ listed the 24 solar terms identical to those of today.

3. Explain the following terms.

1) Twenty-four Solar Terms
2) Clear and Bright

4. Translate the following paragraph into English.

从二十四节气的命名可以看出，节气的划分充分考虑了季节和气候的变化。其中，立春、立夏、立秋、立冬、春分、秋分、夏至、冬至是和春夏秋冬四个季节有关。春分、秋分、夏至、冬至是从天文角度来划分的，反映了太阳高度的变化。而立春、立夏、立秋、立冬则反映了四季的开始。由于中国地域辽阔，各地天气气候差异巨大，因此不同地区的四季变化也有很大差异。

5. Critical thinking and discussion.

As for the specific names of the 24 solar terms, historical and geographical research shows that they first reflected the climate and phenology of the Yellow River reaches during Han Dynasty, and the 24 solar terms are based on this region and its climate. While on the other hand, throughout the vast territory of China, the timing of 24 solar terms used in whole China is the same, what is your understanding?

Section B Farming Activities and Twenty-Four Solar Terms

The 24 solar terms have been the agricultural calendar highly valued by Chinese farmers. Each solar term provides detailed guidelines and timetable for agricultural activities during four seasons and makes it possible for farmers to arrange seeding, field management and harvesting and so on in accordance with seasonal changes. Generally speaking, the rule of "seeding in spring, growing in summer, harvesting in autumn and storing in winter" （春种、夏长、秋收、冬藏） is still followed in terms of 24 solar terms.

"A year's plan starts with spring" （一年之计在于春）, goes the old saying. It reveals that Chinese give top priority to spring because they believe their hard **toil** in spring will bring a good harvest. 3,000 years ago, on the first day of Beginning of Spring, a special ceremony would be held and people would offer sacrifices to **Goumang**[1], the god of spring, who was considered to be in charge of agriculture. By Qing Dynasty, greeting spring had become a **prevalent** folk activity and even government officials would welcome spring in the field near the place of Dongzhimen （东直门） in Beijing. After Beginning of Spring, the day gets longer and the weather warmer.

Goumang

Chapter 3 Twenty-Four Solar Terms

At this important time of the year, temperature, sunlight and rainfall usually start to increase. "Spring begins and rain arrives, so get up early and sleep late"（立春雨水到，早起晚睡觉）, goes the farming proverb. It is to remind farmers that the preparation for annual agricultural work begins.

With the arrival of Rain Water, the weather becomes not so cold when snow and ice **melt**. As the name of this solar term indicates, rain is the most typical feature in this period. According to an old Chinese saying, the rainfall in spring is as precious as oil. The temperature rises **considerably** with more rainfall, which provides favorable growing conditions for winter crops because the rain brings **nutrients** to the soil and plants. Rain Water is also the best time for spring plowing and a precious opportunity for farmers to prepare for selecting seed and applying fertilizer. In northern China, the **precipitation** in this season accounts for a small proportion of annual average rainfall. Therefore, Rain Water is considered as a key period for **irrigation** when the day gets warmer. Insects Awakening is an extremely important time for farmers and is widely taken as the start of the busiest time for farming work. During this period, there is a fast rise in temperature and a marked increase in sunshine in most areas of China, which provides good natural conditions for farming activities. The farming proverb "once Insects Awakening comes, spring plowing never rests"（过了惊蛰节，春耕不停歇）indicates the importance of this solar term to farmers.

Spring Equinox is also an important solar term for farming. Around Spring Equinox, the temperature rises and farm cattle get busy, just as shown in the folk song "In the ninth nine days, the yak starts working"（九九加一九，耕牛遍地走）. It is a **crucial** time for seeding due to melting of frozen soil in most areas and the winter crops enter the growth stage in the spring season. During this period, northern part of China usually suffers from dust and spring drought, therefore it is key time for irrigating the crops. While southern part experiences increase in temperature although there might be late spring coldness and **overcast** weather sometimes, accordingly it is a suitable time to plant rice, corn, and vegetables. In the southern area of the lower reaches of the Yangtze River, there is a custom for farmers to feed the cattle **sticky** rice balls as a reward in order to thank them for their great help in plowing. Meanwhile, people will offer sacrifice to birds, to thank them for bringing signals for farming season and to ask them not to eat grains later in the year. Clear and Bright also marks the time for sowing. From the day on, temperature continues to rise and rainfall increases, making it a vital time for plowing and sowing in the spring. The farmer's proverb says, "melons and beans are seeded around Clear and Bright" (清明前后，种瓜种豆), and trees are also planted during this time.

Grain Rain, as the last term in spring, brings an obvious increase in temperature and rainfall, which has a great influence on the growing of grain crops. The ancient saying "rain brings up the growth of hundreds of grains" indicates that appropriate rainfall will contribute to the **reviving** and growing of crops. Farmers usually seize this period of time to plant rice, corns and beans, expecting the frequent rains for irrigation.

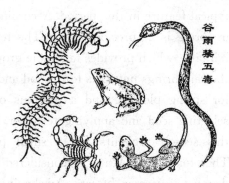
Grain Rain Poster

With the continual increase in temperature, days after Grain Rain enter a peak period of diseases and insect pests, so it is key time to kill the pests in fields for reducing damage to crops. There is also a custom for farmers to post up **Grain Rain posters**[2] to drive away evil force and pray for good luck.

From Beginning of Summer, summer crops enter their final growth stage. As the winter wheat begins its flowering and ripening stage, the date of summer harvest is **tentatively** determined. During Grain Buds, the seeds from summer-harvest crops such as the grain are becoming full but not ripe yet. This time is also a season when plant diseases and pests occur most frequently. Grain in Ear is known as *mangzhong* (芒种) in Chinese. *Mang* (芒) means the ear or the tip of grain, and its pronunciation is the same to the Chinese character "忙", which means busy in English. So the name of this solar term implies double meanings of "when the grain is in the ear" and "the time when farmers are busy". During this period, some awned crops like wheat should be harvested soon and other awned crops like rice seeds can be sown in time. Because people believe that the hot weather in summer after Grain in Ear makes it hard for crops to survive if the deadline for the sowing activities is missed. The agricultural proverbs like "Grain in Ear means both harvest and plant" (芒种芒种，连收带种) and "once the time of Grain in Ear has passed, there will be no use planting anymore" (芒种不种，再种无用) tell the same reason. Therefore, the day of Grain in Ear is regarded as an important dividing point for the time of sowing. During this time, rainfall usually increases much more than before. The regions in the middle and lower reaches of the Yangtze River are about to enter **Plum Rain**[3] season, a good period for growing vegetables and fruits.

Around Summer Solstice, the temperature in most parts of China is pretty high and sunshine is **abundant.** Crops grow fast and need much water. At this time, rainfall is beneficial for the increase in crop **yield.** Usually the rainfall can offer enough water for crop growing in the lower reaches of the Yangtze River and the area between the Yellow River and the Huaihe River (淮河). When Lesser Heat arrives, it will soon stop raining continuously along the Yangtze River. However, the north and north-east of China will enter the rain season. One of the common farming activities in the north is to control flood while in the south is to **relieve drought.** Greater Heat is the time with the highest sunshine and temperature. During this period, crops grow most rapidly because high temperature as well as abundant sunshine and rainfall are favorable for the growth of crops. But natural disasters such as floods, droughts and typhoons also happen during this time. Therefore, it's also the season for farmers to rush to harvest to avoid the damages.

Beginning of Autumn has been of great significance in farming activities since ancient times for it **heralds** the coming of harvest season. As the crops ripen soon, water is still a decisively favorable factor, just as the saying goes, "If it rains on the day of the Start of Autumn, a good harvest is expected" (立秋雨淋淋，遍地是黄金). End of Heat implies that hot summer passes into cool autumn in most parts of China. During this period, in North China the crops ripen quickly due to great temperature differences between day and night. While, in South China, it is the time to harvest middle-season rice. When White Dew comes, farmers are alert to the autumn rain which may cause damage to the crops, and at the same time they make preparations for rain water storage. As the weather becomes colder after White Dew, there comes the period to take precautions against cold and pests.

Around Autumn Equinox, farmers all over the country begin to experience *sanmang* (三忙) which means they are extremely busy harvesting, plowing, and sowing. On the one hand, farmers are on the wing harvesting cotton and late wheat. On the other hand, they are busy plowing and selecting the seeds of wheat, barley and broad beans to prepare for sowing. During this period, the farmers should harvest the crops as early as possible in case of the continuous rain as well as the coming frost. And they should sow the winter crops as early as possible so as to **breed** strong **seedlings** for **overwintering**, which may lay a foundation for high yield in the next year. The coming of Cold Dew indicates that farmers should seize the last chance to harvest and plow. Otherwise, the crops will be damaged by frost later. During First Frost, frost falls frequently, which may result in crops cut. Consequently, some measures should be immediately taken against it.

Beginning of Winter marks the coming of winter and crops enter overwintering. It is the best time for wheat planting in South China, while in North China, the farmers should irrigate wheat fields to resist frost. During Light Snow, most parts of the country begin to prepare for the severe cold. Farmers are busy storing the harvested vegetables such as Chinese cabbage, carrots and potatoes. With the coming of Heavy Snow, it snows more heavily in northern China, which plays a very important role in farming. On the one hand, snow creates a good overwintering environment for winter crops. On the other hand, it increases the soil **moisture** necessary for crops growth in the spring. In addition, compared with rain water, snow water can better fertilize the soil as it contains **nitrides** five times as much as ordinary rain water. And the importance of snow is vividly described in farmers' proverb "three layers of snow on wheat this year, then a sound sleep on bread next year" (冬天麦盖三层被，来年枕着馒头睡). From Winter Solstice, the weather becomes much colder and it is no longer a suitable period for large-scale farming activities, but a period for farmers to manage the field of winter crops. When Lesser Cold and Greater Cold come, the temperature will drop to the lowest in a year. This period is the most important time for fertilization, cold prevention and **maintenance** of irrigation equipment. Besides, the cold temperature serves as the natural **pesticide** in the coming year.

Since ancient times, the system of the 24 solar terms has been a significant reference for farmers to undertake farming activities, which has had a positive impact on ancient agricultural developments. In the current time of advanced science and technology, its role has **diminished** because farmers have **innovated** their farming concepts and methods to cultivate and harvest crops. Despite the decreasing dependence on the 24 solar terms, farmers today still frequently **consult** the 24 solar terms for farming work in different areas across China based on their specific local conditions. The system of the 24 solar terms reflects the experience and wisdom of the ancient Chinese and it is China's precious agricultural **heritage**.

New words

toil [tɔɪl] *n.* Hard or exhausting work. 辛苦的工作

prevalent [ˈprevələnt] *adj.* A condition, practice, or belief that is prevalent is common. 盛行的；普遍存在的

melt [melt] *v.* When a solid substance melts or when you melt it, it changes to a liquid, usually because it has been heated. 使融化；融化

considerably [kənˈsɪdərəbli] *adv.* 相当大地；相当多地；显著地

nutrient [ˈnjuːtriənt] *n.* Nutrients are substances that help plants and animals to grow. 营养物

precipitation [prɪˌsɪpɪˈteɪʃn] *n.* Precipitation is rain, snow, or hail. 降水

irrigation [ˌɪrɪˈɡeɪʃn] *n.* 灌溉

crucial [ˈkruːʃl] *adj.* If you describe something as crucial, you mean it is extremely important. 至关重要的

overcast [ˈəʊvəkɑːst] *adj.* If it is overcast, or if the sky or the day is overcast, the sky is completely covered with cloud and there is not much light. 阴天的

sticky [ˈstɪki] *adj.* A sticky substance is soft, or thick and liquid, and can stick to other things. 黏的

revive [rɪˈvaɪv] *v.* When something is revived or when it revives, it becomes active again. 苏醒；复活

tentatively [ˈtentətɪvli] *adv.* 暂时地；试验性地

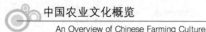

abundant [əˈbʌndənt] *adj.* Something that is abundant is present in large quantities. 丰富的

yield [jiːld] *n.* Yield is the amount of crops that are produced. 产量

relieve [rɪˈliːv] *v.* If something relieves an unpleasant feeling or situation, it makes it less unpleasant or causes it to disappear completely. 缓解；减轻；解除

drought [draʊt] *n.* A drought is a long period of time during which no rain falls. 干旱

herald [ˈherəld] *v.* Something that heralds a future event or situation is a sign that it is going to happen or appear. 预示…的来临

breed [briːd] *v.* If you breed animals or plants, you keep them for the purpose of producing more animals or plants with particular qualities, in a controlled way. 养殖

seedling [ˈsiːdlɪŋ] *n.* A seedling is a young plant that has been grown from a seed. 幼苗

overwinter [ˌəʊvəˈwɪntə] *v.* To spend winter (in or at a particular place)（在某地）过冬

moisture [ˈmɔɪstʃə] *n.* Moisture is tiny drops of water in the air, on a surface, or in the ground. 潮气；水分

nitride [ˈnaɪtraɪd] *n.* 氮化物

maintenance [ˈmeɪntənəns] *n.* The maintenance of a building, vehicle, road, or machine is the process of keeping it in good condition by regularly checking it and repairing it when necessary. 检修

pesticide [ˈpestɪsaɪd] *n.* Pesticides are chemicals that farmers put on their crops to kill harmful insects. 杀虫剂

diminish [dɪˈmɪnɪʃ] *v.* When something diminishes, or when something diminishes it, it becomes reduced in size, importance, or intensity. 使减少；变少

innovate [ˈɪnəveɪt] *v.* To innovate means to introduce changes and new ideas in the way something is done or made. 创新；革新

consult [kənˈsʌlt] *v.* If you consult a book or a map, you look in it in order to find some information. 查阅

Phrases and expressions

give top priority to　优先考虑
in charge of　负责，主管
account for　对…做出解释；说明…原因
be alert to　对…警惕；对…警觉
be on the wing　忙于…
take precautions against　采取预防措施
lay a foundation for　奠定基础

Cultural Notes

1. Goumang（句芒）is the spring god with a bird face according to the legend. People believe that Goumang is a young man holding a whip, dressed in red, wearing a straw hat and straw sandals. When worshiping Goumang, a straw man will be made in fixed size, who is 3 feet and 6 inches high, symbolizing the 365 days a year and a bumper grain harvest. The whip in his hand is 2 feet and 4 inches long, representing the 24 solar terms in one year.

2. Grain Rain poster is a kind of New Year painting and it is a combination of Taoism and folk customs in China. It is usually carved with images such as chickens catching scorpions and Taoist masters removing five poisons, which shows people's desire for harvest and peace. During Grain Rain, as the temperature rises, pests will enter the high reproduction period. Besides, in the past the building materials of the house gave the pests hiding space, which may cause the risk of stings easily. As a result, when Grain Rain comes, farmers tend to post Grain Rain poster in order to reduce damage to crops and people. This custom is very popular in Shandong, Shanxi Province.

3. Plum Rain generally begins in mid-June and ends in mid-July although the timing varies from year to year and from place to place in some parts of China. It refers to the long period of continuous rainy or cloudy weather. The rain is called Plum Rain as it overlaps with the time when plums ripen. During Plum Rain season, high humidity and high temperatures make clothes go moldy which is why it is also called "mold rain season".

Exercises

1. Read the following statements and try to decide whether it is true or false according to your understanding.

____ 1) The solar term Rain Water indicates that appropriate rainfall during this period will contribute to the reviving and growing of winter crops as well as the sowing and seeding of crops in spring.

____ 2) As Insects Awakening comes, farm work starts and both the farmers and the cattle start to become busy. It is a custom for farmers to reward cattle with sticky rice balls to express their gratefulness.

____ 3) During Grain in Ear, the regions in the middle and lower reaches of the Yangtze River are about to enter Plum Rain season, which is a good period for growing vegetables and fruits.

____ 4) Start of Autumn is an important solar term for farmers for it is time to gather crops. There is a saying: "If it rains on the day of Start of Autumn, a good harvest is expected."

____ 5) It is safe to say that the winter is not a good time for farming due to its coldness and it is more harmful than beneficial.

2. Fill in the blanks with the information you learn from the text.

1) 3,000 years ago, on the first day of Start of Spring, people in China began holding a special ceremony and they offered sacrifices to _____.

2) During Spring Equinox, there is a custom for farmers to feed

_____ the sticky rice balls as a reward in order to thank them for their great help in plowing.

3) The name of _____ implies a double meaning of "when the grain is in the ear" and "the time when farmers are busy".

4) When _____ comes, farmers are alert to the autumn rain which may cause damage to the crops, and at the same time they make preparations for rain water storage.

5) Beginning of Winter marks the coming of winter and crops enter overwintering. It is the best time for _____ in South China, while in North China, the farmers should irrigate wheat fields to _____.

3. Explain the following terms.

1) *Sanmang*
2) Plum Rain

4. Translate the following paragraph into English.

二十四节气是古代中国订立的一种用来指导农事的历法。在人类从事农业生产的早期，就发现播种收获等农事活动与季节有着非常密切的关系。通过不断观察总结，到了秦汉时期，用以确定农时、指导农业生产的二十四节气完整建立。二十四节气中部分节气甚至直接用农时命名，如"小满""芒种"。二十四节气客观地反映了中华民族发源地——黄河流域地区的季节更替和气候变化状况，是古代中国农业文明的具体表现。

5. Critical thinking and discussion.

The 24 solar terms indicate the ancient Chinese people's experience in agricultural production and reflect their thought of unity of Heaven and Man. They not only played a very important role in guiding farming activities in ancient time, but also guide Chinese people's farming in modern society. What roles do you think they have in modern agriculture? And what farming wisdom do you think they reflect?

Section C Dietary Customs and Twenty-Four Solar Terms

The 24 solar terms are considered, besides guiding farming, to have a great effect on people's **dietary** habits. People can adapt their diets to the unique weather changes of each solar term. This can help to relieve fatigue, keep fit and stay young from the perspective of health preservation. In addition, many dietary customs have been formed based on certain solar terms over the long history of development and certain food will be taken usually for specific wishes.

In Spring, the nature comes alive and all living things are **flourishing**. Similar to the crops, human bodies also start a new round of growth. It is undoubted that spring is the best time for people to wake up the body and enhance **immunity**. Meanwhile, spring is also time to nourish liver, for the germs and viruses tend to become active in such warm weather and often lead to liver-related diseases. As a result, People are strongly suggested to take liver-nourishing food like **fermented** soya beans, spring onions, caraways and peanuts.

Qingtuan

With the coming of Beginning of Spring, people prefer to eat spring pancakes, spring rolls or turnips to welcome the coming of spring, which is called *yaochun* (咬春). It is well known that the wet weather is unfavorable to people's spleen and stomach. During Rain Water, as the rainfall increases, people are recommended to take more dampness-dispelling foods like yams, millet, carrots and crucian carp to help protect the digestive system. As it gets warm and dry, there comes Insects Awakening, a high-occurrence period for diseases such as flu, chickenpox and oral herpes, people are suggested to take light foods instead of greasy ones. On Spring Equinox, it is customary for people to have *chuncai* (春菜), the seasonal vegetables in spring. The Chinese medical classic *Huangdi Neijing* suggests that people have seasonal food to keep healthy. The period of Clear and Bright comes with two important days, **Hanshi Festival**[1] and Qingming Festival. On the former day, there is a traditional **taboo** to make fire and people are only allowed to take cold food. On the latter day, people in different places have different foods. In Shandong Province, people eat eggs and *lengbobo* (冷饽饽), a cold staple food made of flour. While in Shanghai, people prefer *qingtuan*（青团）, a glutinous rice ball with sweet bean paste. At the end of spring, there comes Grain Rain, the northern people love to eat Chinese toon（香椿）because it is fresh and nutritious at this moment, just as the old saying goes "toon before the rain is as tender as silk"（雨前香椿嫩如丝）. The southern people pick some new tea leaves for drink. Ancient legends say drinking new tea on Grain Rain helps clear *huo*（火）, an internal heat.

In summer, it is extremely critical to nourish heart, because the high heat at this period usually has a bad effect on the heart, so summer is a high-occurrence season for heart-related diseases. On the one hand, it is advisable to eat foods rich in vitamins and cool in nature to clear away heat such as bitter gourds, water melons, musk melons and mung beans. On the other hand, it is good for people to take the red-color food like Chinese dates, red beans and cherries, which is considered to help lower blood **lipid**.

On the very day of Beginning of Summer, people in some regions prefer to eat eggs for the wish of good health, for they believe a round egg symbolizes a happy life. In addition, the southerners take "colorful rice", called *lixiafan*（立夏饭）, which is made from many different kinds of beans mixed with rice. Grain Buds is a perfect time to eat fish and shrimp

Lixiafan

which are fresh and fat in this period because of the increasing rainfall and the rising water level. It is also time to take the sowthistle（苦菜）, which helps cool and **detoxify** body. When Grain in Ear falls, there comes the Plum Rain Season（梅雨季）in the south and the plums are going to ripen. They greatly benefit people for being rich in minerals and natural organic acids. As the temperature rises during Summer Solstice, people often feel tired and suffer from headache because of the **insufficient** blood to the brain which is caused by the loss of body fluid due to sweating. Therefore, it is necessary for people to drink much water. In dining, people are recommended to eat more sour food which may level the stomach acids to kill the bacteria which can cause stomach and intestinal diseases. Besides, on the day of Summer Solstice, there is a tradition for people to take cold noodles in many regions of China, such as Shandong Province, as the saying goes "take noodles on Summer Solstice and *jiaozi* on Winter Solstice". Lesser Heat and Greater Heat are the periods for the melon. One custom in Nanjing has something to do with enjoying small melons like hami melons on the day of Lesser Heat and having big melons such as watermelons on the day of Greater Heat. During these two hot periods, people had better take less food with hot nature such as lamb, pork and onions.

Autumn is usually marked by dryness, hence the term "autumn dryness". When the humidity in the air drops, people's lung, skin and hair are easily affected. People may feel skin-stretched and thirsty, and easily cough. Therefore autumn is the best time for people to nourish lung. In dining, they are recommended to eat more fresh vegetables and fruits to moisten body, especially the lung. For instance, the uncooked and cooked pears are the top choices. The uncooked ones help to clear away heat while the steamed ones help to nourish lung.

When Beginning of Autumn comes, it is popular for people to have a variety of delicious food, especially meat, to regain the lost weight in summer, called *tieqiubiao* (贴秋膘). During End of Heat, people often feel weary, called "autumn weariness". It is a sign that the human body needs rest, since people have consumed excessive energy in summer. People should maintain a light diet, for the greasy food may produce **acidic** substances in the body which makes people feel more tired. And people are highly recommended to take more fresh vegetables and fruits which are **alkaline** substances. These help to balance acidic substances and to relieve fatigue effectively. On the day of White Dew, it is custom for people to eat sweet potatoes in some places nowadays. And White Dew is also time when fruits like grapes, pears and longans become widely available. Eating these seasonal fruits can help clear one's internal heat and moisten lung. On Autumn Equinox, the southerners customarily have *qiucai* (秋菜), the seasonal vegetables in autumn, and take *qiutang* (秋汤), a soup made of fish and vegetables. A popular saying about the soup goes that "drinking the soup to clear the liver and intestines, thus the whole family will be safe and healthy" (秋汤灌脏，洗涤肝肠。阖家老少，平安健康。). During Cold Dew, people in many regions have the custom of drinking chrysanthemum wine（菊花酒）to resist dryness. Meanwhile, Cold Dew is also time to harvest hawthorn which contains the antioxidants（抗氧化剂）helping to strengthen blood vessels. First Frost is an important period for people to resist cold. People can take some persimmons to keep away from cold and protect their bones. Among the seasonal vegetables, turnip is a top choice, which may help relieve stomach upset.

Qiutang

In winter, people's **metabolism** gets slower as the weather gets colder, so they must rely more on the **motivity** of kidney to resist cold. With the decline of kidney function, people will always have a bad memory, poor **appetite**, and even lose sleep. Therefore, this period is the proper time to nourish kidney by taking mutton, beef, chicken and bone soup.

There is a saying about Beginning of Winter in the north that goes "have *jiaozi* on Start of Winter, or your ears will be **frostbitten**". This custom derives from a story about **Zhang Zhongjing**[2]. He saved many people in Henan Province from a typhoid around Start of Winter in the late Eastern Han Dynasty. He cooked mutton with hot peppers and herbs to dispel cold

Ciba

and to prevent people's ears from being frostbitten. He wrapped these ingredients into a **dough** and shaped them into an ear. That was the earliest *jiaozi*. And eating *jiaozi* on Start of Winter graduated into a widely-accepted practice in China. With a rapid drop in temperature during Light Snow and Heavy Snow, there comes the time for people to make preserved food, such as the smoked meat. In the south, local people have the custom of eating *ciba* （糍粑）made by glutinous rice with sugar. Winter Solstice is special for Chinese since it is close to the Spring Festival. On that day, *jiaozi* is a must in the north, and wonton（馄饨）is another option for people. In the south, people celebrate this day by eating *tangyuan*, the sticky rice balls with fillings, which symbolize "reunion" because of the round shape. During Lesser Cold and Greater Cold, the coldest days of the year, it is a good choice for people to take some food with hot nature to nourish and warm body. The northerners prefer to take mutton soup, while the southerners have the sticky rice mixed with some fried preserved pork, sausages and peanuts.

Since the ancient time, Chinese people began to realize that the changes of human bodies and certain diseases are closely related to the 24 solar terms. It is undoubted that the system of the 24 solar terms plays a very important role in guiding people's dietary habits. Although, people's dietary customs vary from region to region under its influence, they generally follow the same principle from the perspective of health preservation, to nourish liver in spring, heart in summer, lung in autumn and kidney in winter. All these are the good examples to show the long-held idea by the Chinese people that man and nature are closely related to each other, and man should keep harmony with nature.

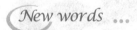

dietary [ˈdaɪətərɪ] *adj.* You can use dietary to describe anything that concerns a person's diet. 饮食的

flourish [ˈflʌrɪʃ] *v.* If a plant or animal flourishes, it grows well or is healthy because the conditions are right for it. （动植物因环境适宜而）旺盛

immunity [ɪˈmjuːnɪtɪ] *n.* 免疫力

ferment [fəˈment] *v.* If a food, drink, or other natural substance ferments, or if it is fermented, a chemical change takes place in it so that alcohol is produced. 使发酵；发酵

taboo [təˈbuː] *n.* A taboo against a subject or activity is a social custom to avoid doing that activity or talking about that subject, because people find them embarrassing or offensive. 禁忌

lipid [ˈlɪpɪd] *n.* 脂质；油脂

detoxify [ˌdiːˈtɒksɪfɪ] *v.* If you detoxify, or if something detoxifies your body, you do something to remove poisonous or harmful substances from your body. （使）解毒

insufficient [ˌɪnsəˈfɪʃənt] *adj.* Something that is insufficient is not large enough in amount or degree for a particular purpose. 不充分的

acidic [əˈsɪdɪk] *adj.* Something that is alkaline contains an alkali or has a pH value of less than 7. 酸性的

alkaline [ˈælkəlaɪn] *adj.* Something that is alkaline contains an alkali or has a pH value of more than 7. 碱性的

metabolism [məˈtæbəlɪzəm] *n.* 新陈代谢

motivity [məʊˈtɪvətɪ] *n.* The power of moving or of initiating motion 动力；动力

appetite [ˈæpətaɪt] *adj.* Your appetite is your desire to eat. 胃口

frostbite [ˈfrɒstˌbaɪt] *v.* Frostbite is a condition in which parts of your body, such as your fingers or toes, become seriously damaged as a result of being very cold. 冻伤

dough [dəʊ] *n.* Dough is a fairly firm mixture of flour, water, and sometimes also fat and sugar. It can be cooked to make bread, pastry, and biscuits. 面团

Phrases and expressions

have something to do with　　跟某事有关
derive from　　源出，来自
graduate into　　渐渐变为

Cultural Notes

1. Hanshi Festival（寒食节）or the Cold Food Festival is a traditional Chinese holiday celebrated the day before Clear and Bright. Legend has it that during the Spring and Autumn Prince Chong'er（重耳）of the State of Jin endured many hardships while he was exiled from his home state. Once, in order to help the Prince who was tormented by hunger, Jie Zitui（介子推）cut off the flesh from his thigh and offered it to the Prince. When Chong'er became Duke Wen of Jin, he did not reward him. Instead, he killed Jie and his mother in a fire. Later, Duke Wen, filled with regret, ordered that using fire on the anniversary of Jie's death was forbidden and all food was to be consumed cold.

2. Zhang Zhongjing（张仲景）was born in the late Eastern Han dynasty and was a famous medical scientist, later honored as the medical sage（医圣）. He extensively collected the medical prescriptions, and wrote a masterpiece, named *The Theory of Typhoid Fever*（《伤寒杂病论》）, in which he recorded a large number of effective prescriptions. This book becomes indispensable for scholars to study traditional Chinese Medicine and it is widely brought to the attention of the medical students and clinical doctors.

Exercises

1. Read the following statements and try to decide whether it is true or false according to your understanding.

____ 1) When the Spring Equinox is coming, there comes Hanshi Festival.

____ 2) During Grain in Ear, the plums are going to ripen, which are rich in minerals and natural organic acids.

____ 3) Since the weather is hot during Lesser Heat and Greater Heat, people are encouraged to take more lamb, pork, hot peppers, onions and ginger.

____ 4) On Beginning of Winter, there is a saying that goes "Eat dumplings on Start of Winter Day, or your ears will be frostbitten".

____ 5) Beginning of Autumn is a key time to moisten the heart.

2. Fill in the blanks with the information you learn from the text.

1) When the season of spring comes, in many parts of China, people observe the custom of "_____" on the first day of Beginning of Spring and they usually eat spring pancakes, spring rolls or turnips to welcome the coming of spring.

2) Insects Awakening is a high-occurrence period for diseases such as flu, chickenpox and oral herpes, people are suggested to take _____ foods instead of _____ ones.

3) There is a saying in Shandong province which goes, "eat _____ on Winter Solstice and eat _____ on Summer Solstice."

4) As the weather becomes cool in autumn, many people will feel weary, which is called "_____".

5) On the day of Beginning of Summer, people like to eat _____ as a prayer for good health, for they believe it symbolizes a happy life.

3. Explain the following terms.

1) Hanshi Festival

2) *Tieqiubiao*

4. Translate the following paragraph into English.

从养生的角度来看，人类机体的变化、疾病的发生都与二十四节气紧密相连。因此养生学家认为，人们可以通过季节变化来进行养生保健活动。在各个时节选择不同的食物和出行方式来满足自身体内需求，让身体随着时节的变化而改变，以达到强身健体、延年益寿的目的。

5. Critical thinking and discussion.

The 24 solar terms have a long history of thousands of years. They are considered to have a great effect on people's dietary habits. People can adapt their diets to the unique weather changes of each solar term in order to keep fit. However, in the modern society some people hold the view that one should take the seasonal food according to the weather changes of each solar term in order to keep healthy, while others hold the view that one don't need to follow these rules and one should take any food that they like. What's your point of view?

Chapter 4
Typical Chinese Food Crops and Farming Tools

Section A Typical Food Crops in China

As one of the cradles of world civilizations, China has a long history of crop cultivation. Archaeological evidence indicates that ancient Chinese people in Yellow River Valley and Yangtze River Valley started **cultivating** crops such as millet and rice around 10,000 years ago. Besides, food crops like corn and sweet potatoes from other regions or countries have also been introduced into China, enriching its crop variety. Both the native and foreign food crops have contributed a lot to feeding the ever-increasing Chinese population.

Five Grains

In Chinese legends, Shennong, the legendary emperor of China, domesticated crops like wheat, rice, millet, bean, and soybean and instructed people to plant them. These crops are often referred to as Five Grains. Therefore, Shennong was also known as "The Emperor of Five Grains". In fact, all of the five crops were depicted in **Book of Songs**[1], like "Large rat, large rat, eat no more our millet…", which proves that the crops were already important to people in Pre-Qin periods. What's more important is the increasing archaeological findings of their early cultivation in China. Even though there are some other versions of Five Grains, the one mentioned above is much widely-accepted and considered as the earliest crops planted in China.

Millet, topping the list of Five Grains, is recognized as the most important food crop in ancient China. About 7,000 to 8,000 years ago, millet was planted along the Yellow River Valley, which marked the beginning of primitive agriculture in northern China. The millet **remains** found in northern China at the historical sites of Banpo village (半坡村), Shaanxi Province and

Peiligang (裴李岗), Henan Province, provide reliable evidence for its early cultivation in China.

Millet cultivation provided **staple** food for ancient Chinese and **nourished** the Chinese civilization as well. With its dominant position in the grains, some sacred meaning was added to it, and it became the **embodiment** of **God of Grains**, or *ji* (稷) in Chinese. Another god that was in charge of harvest was the God of Land, which was called *she* (社) in Chinese. Every year, both the gods were offered sacrifices by ancient emperors in order to pray for peace and harvest. In Zhou Dynasty, *she* and *ji* started to be used as a fixed phrase *sheji*, symbolizing China the nation.

Millet had been cultivated as a staple crop in China till the Northern and Southern Dynasties. Later, with the increase of rice and wheat planting, millet planting was gradually reduced.

Broom-corn millet (黍) was an annual grass crop domesticated along the Yellow River Valley in the same period of millet. It served as staple food and raw materials for making liquor in ancient times as well. It was almost as important as millet in the grain crops in Shang and Zhou Dynasties.

The remains of broom-corn millet were found in many sites of northern China, mainly on the middle **reaches** of Yellow River. The earliest remains were discovered in Dadiwan **ruins** (大地湾遗址), Gansu Province, about 8,200 years ago. In either **Oracle Bone Script**[2] or *Book of Songs*, one could find records about broom-corn millet and millet more than other crops, which reveal their important position in ancient people's life.

However, the yield of broom-corn millet was low and the taste was not quite appealing to people. Besides, rice and wheat started to be cultivated extensively in Song Dynasty. Gradually, broom-corn millet was planted only in some cold areas in the north of China.

Wheat was an annual grass crop found to be cultivated in **Tarim Basin** and Mount Tianshan in Xinjiang about 4,000 years ago. Scholars tend to believe that wheat was first cultivated in the regions of the **Fertile Crescent**[3] and later introduced to China.

The planting of wheat has gone through several stages. During the Warring States Period, wheat planting was expanded due to the invention of iron farming tools and the agricultural policies encouraging people to plant

Chapter 4 Typical Chinese Food Crops and Farming Tools

wheat. Not surprisingly, in Han Dynasty wheat yield **surpassed** broom-corn millet, only second to millet.

In Wei, Jin, Southern and Northern Dynasties, the increase of wheat planting continued mainly due to large population migration. Wars in the north drove lots of people to the south, together with them the farming techniques and tools. Accordingly, the crops cultivated in the north started to be planted in the south. The second large population migration to the south happened in the early Southern Song Dynasty. The growth in population led to a huge demand for wheat. To meet the challenge, even larger farming land got cultivated for wheat planting, new agriculture tools were invented, and farming skills were developed, such as crop rotation. In the Southern Song Dynasty, the yield of wheat surpassed that of millet.

Rice, an annual grass herbal crop, was **domesticated** from the wild rice growing along the Yangtze River Valley about 10,000 years ago. At Hemudu site（河姆渡遗址）, one of the earliest New Stone Age locations along the lower reaches of the Yangtze River, piles of rice grains, husks, stalks and leaves were found. This **verifies** that China was one of the key areas in the world where rice originated.

Despite a long history of cultivation, the yield of rice remained much less than that of millet till Han Dynasty. The main reason was that the south was **sparsely** populated then. Later, people in the north were compelled to move to the south because of the social unrest in Song Dynasty, which resulted in the population in the Yangtze River Valley **outnumbering** that in the Yellow River Valley. The population growth stimulated a high demand for food, which in turn accelerated the development of farming skills and tools. Eventually the yield of rice exceeded that of any other grain crops in this period. Since then, the Yangtze River Valley become the economic center of China.

Soybean was domesticated between 7000 and 6600 BC in China. It became a main crop by Zhou Dynasty in the north of China and was spread to the south in Han Dynasty. Soybean was quite popular with ancient Chinese. It was processed into a variety of soybean-related seasonings and foods, such as sauce, bean paste and tofu. Today, Soybean is playing an important role as one of the important oil crops in people's daily life in the world.

Corns and Sweet Potatoes

Besides growing new varieties of local food crops, Chinese people planted food crops from other regions and countries. Some came from **Western Region**[4] with the opening of Silk Road, some from Central Asia and South Asia in the frequent foreign exchanges in Tang Dynasty and some from America after the American continent was discovered. All the crops have contributed a lot in feeding the large population of China. Among them, corn and sweet potato are the typical representatives.

Corn, originating in Central America or South America, was introduced to China in Ming Dynasty. Li Shizhen, a medical expert of Ming Dynasty, recorded this crop in his ***Compendium of Materia Medica***. Being high-yield, easy to be planted and tolerant of drought and other extreme weather conditions, the crop was widely planted in the country in the 17th century. It was recorded that corn-growing encouraged population growth. Today, it is still cultivated in large areas in China, but mainly consumed as animal **fodder**.

Sweet potato is thought to be firstly planted in either Central America or South America and was introduced into China via the Philippines into Fujian Province in Ming Dynasty. Sweet potato is high-yielding and resistant to bad weather and diseases. In addition to serving as a staple food, it can be made into wine, sugar, starch noodles and other products. Xu Guangqi (徐光启), a great agricultural scientist in Ming Dynasty, had once written a book to introduce its benefits and planting methods. In the early years of Qing Dynasty, sweet potato was spread quickly across the country. Historically, the planting of sweet potatoes **spurred** the population explosion.

All the above-mentioned crops have played an important role in driving the social progress, especially those originated in China. The cultivation of those crops marked the beginning of ancient agriculture and the end of a life roaming from one place to another to gather wild grains for food. Thanks to the crops, ancient Chinese settled down and established one of the earliest civilizations in the world.

New words

cultivate [ˈkʌltɪveɪt] *v.* It means you prepare land and grow crops on it. 种植

remains [rɪˈmeɪnz] *n.* Historical remains are things that have been found from an earlier period of history, usually buried in the ground, for example parts of buildings and pieces of pottery. （历史）遗迹；残迹

staple [ˈsteɪp(ə)l] *adj.* A staple food, product, or activity is one that is basic and important in people's everyday lives. 基本的；主要的（食物、产品、活动）

nourish [ˈnʌrɪʃ] *v.* To nourish a person, animal, or plant means to provide them with the food that is necessary for life, growth, and good health. 为…提供营养

embodiment [ɪmˈbɒdɪmənt] *n.* If you say that someone or something is the embodiment of a quality or idea, you mean that that is their most noticeable characteristic or the basis of all they do. 集中体现；化身

reaches [ˈriːtʃɪz] *n.* The upper, middle, or lower reaches of a river are parts of a river. The upper reaches are nearer to the river's source and the lower reaches are nearer to the sea into which it flows. 河段；流域

ruin [ˈruːɪn] *n.* The ruins of something are the parts of it that remain after it has been severely damaged or weakened. 废墟；遗迹

surpass [səˈpɑːs] *v.* If one person or thing surpasses another, the first is better than, or has more of a particular quality than, the second. 优于；超过

domesticate [dəˈmestɪkeɪt] *v.* When people domesticate wild animals or plants, they bring them under control and use them to produce food or as pets. 驯养（动物）；培养（野生植物）

verify [ˈverɪfaɪ] *v.* If you verify something, you check that it is true by careful examination or investigation. 核实

sparse [spɑːs] *adj.* Something that is sparse is small in number or amount and spread out over an area. 稀疏的

outnumber [aʊtˈnʌmbə] *v.* If one group of people or things outnumbers another, the first group has more people or things in it than the second group. 在数量上超过

soybean [ˈsɔɪbiːn] *n.* Soybeans are beans that can be eaten or used to make flour, oil, or soya sauce. 大豆

fodder [ˈfɒdə] *n.* Fodder is food that is given to cows, horses, and other animals. 饲料

spur [spɜː] *v.* If something spurs a change or event, it makes it happen faster or sooner. 使更快发生；加速

Phrases and expressions

in charge of　负责；掌管
appeal to　吸引人
second to　次于，仅次于
resistant to　对…有抵抗力的；耐…的

Proper names

Five Grains　五谷
God of Grains　谷神
God of Land　土地神
Tarim Basin　塔里木盆地
Compendium of Materia Medica　《本草纲目》

Chapter 4 Typical Chinese Food Crops and Farming Tools

Cultural Notes

1. Book of Songs, also *Shijing* (《诗经》) or *Shih-ching*, translated variously as the *Classic of Poetry*, *Book of Odes*, or simply known as *the Odes* or *Poetry*, is the oldest existing collection of Chinese poetry, comprising 305 works dating from the 11th to 7th century BC. It is one of the "Five Classics" traditionally said to have been compiled by Confucius, and has been studied and memorized by scholars in China and neighboring countries for over two millennia. Since Qing dynasty, its rhyme patterns have also been analyzed in the study of old Chinese phonology.

2. Oracle Bone Script （甲骨文） refers to the form of Chinese characters engraved on animal bones or turtle shells used in divination by fire （火占卜） in Shang Dynasty, which is the earliest known form of Chinese writing. The vast majority were found at the *Yinxu* （殷墟） site (in modern Anyang, Henan Province).

3. Fertile Crescent （肥沃地带） also known as the "cradle of civilization", is a crescent-shaped region where agriculture and early human civilizations like the Sumer and Ancient Egypt flourished thanks to the surrounding Nile, Euphrates and Tigris rivers.

4. Western Regions （西域） referred to the regions west of Yumen Pass （玉门关） between the 3rd century BC to the 8th century AD. Specifically, it was used more often to refer to Central Asia, or more generally other regions to the west of China. Because of its strategic location astride the Silk Road, the Western Regions have been historically significant since at least the 3rd century BC.

Exercises

1. Read the following statements and try to decide whether it is true or false according to your understanding.

____ 1) According to Chinese mythology, Five Grains were discovered by

Shennong, the legendary emperor of China.

____ 2) Millet originated in the Yangtze River valley in China about 7,000 to 8,000 years ago.

____ 3) Millet and broomcorn millet are the oldest cereal crops in China and in the world.

____ 4) Wheat was first cultivated in the regions of the Fertile Crescent.

____ 5) In Han Dynasty, the yields of rice surpassed any other cereal crops in China.

2. Fill in the blanks with the information you learn from the text.

1) The cultivation of _____ marked the beginning of primitive agriculture in northern China.

2) Wheat was spread to the southern region in China because _____.

3) _____ was domesticated along the Yangtze River Valley about 10,000 years ago but till Southern Song Dynasty its yield came to exceed that of any other grain crops.

4) Before Tang Dynasty, the cultural and economic center of China was in the _____, while in Song Dynasty, the center was moved to _____.

5) Chinese people planted food crops from other regions and countries. For example, the crops brought in from _____ with the opening of "silk road".

3. Explain the following terms.

1) Five Grains
2) *Sheji*（社稷）

4. Translate the following paragraph into English.

"五谷",古代有多种不同说法,最主要的有两种:一种指稻、黍、稷、麦、菽;另一种指麻、黍、稷、麦、菽。两者的区别是:前者有稻无麻,后者有麻无稻。古代经济文化中心在黄河流域,稻的主要产地在南方,而北方种稻较少,所以"五谷"中最初无稻。但随着社会经济和农业生产的发展,五谷的概念在不断演变着,现在所谓"五谷",实际只是粮食作物的总称,或者泛指粮食作物罢了。

5. Critical thinking and discussion.

China is regarded by many agriculturists and botanists （植物学家）as a very important birthplace of numerous cultivated plants. So, some people think the Chinese traditional culture is deeply rooted in its agricultural civilization. Do you agree on this statement? Why or why not?

Section B Traditional Chinese Farming tools

Farming tools are the earliest labor-saving devices in the history of human civilization, generally considered as the mother of all tools. They were invented in agricultural activities for soil preparing, sowing, irrigating, harvesting and other purposes, and greatly improved working efficiency. Moreover, many of the tools took a lead in the world at that time.

Soil Preparing Tool—Plough

Soil preparation is the first stage in the cultivation of land. People need to break **topsoil** and turn it over. In this way, the weeds can be killed and oxygen and water can be easily carried to the roots of crops.

In the early period of **New Stone Age**[1], the tool for soil preparation was *lei* （耒）, a pointed wooden stick. Later, spades made of stones and animal bones were fastened to *lei*, which was named *si* （耜）or *leisi* （耒耜）, to increase its **durability** and efficiency. And later, *leisi* evolved into tools with straight handles like shovel and with curved handles like plough.

In Spring and Autumn Period, with the development of iron smelting technology the stone part of *leisi* was replaced by iron, which made it much sharper. The wooden frame of the tool was designed into a shape to be powered by animals. The primitive plough came into being.

During Han Dynasty the tool evolved into an iron **moldboard** plough （铁铧犁）. **Plowshares** were made into triangular shape to easily cut soil and U-shaped moldboards were fixed to the back of plough to turn over and crush the earth. The combination allowed people to cut and crush soil at the same time, productivity being improved greatly. The technology was introduced to Europe about 2,000 years later.

By Tang Dynasty, the curved-**shaft** plough (曲辕犁) appeared. Greatly improved from earlier types of ploughs, this kind of plough could turn both left and right easily, and even make a U-turn. In the Frescoes (壁画) at Dunhuang Grottoes (敦煌石窟) the plough at work was portrayed. As it was firstly used in Jiangdong (江东), the modern Suzhou area, it was also called Jiangdong plough (江东犁). The curved-shaft plough was widely popular during Song and Yuan Dynasties.

Curved-shaft Plough in Tang Dynasty

The invention of iron ploughs is considered as a significant breakthrough in making farming tools. In contrast to the **intermittent** movement of previous tools, ploughs could cut and crush the soil in a continuous way, and thus people could prepare a field much faster. Even the modern ploughs drawn by tractors are similar to the traditional ones in structure.

Sowing Tool—Seed Plough

Seed sowing is another important procedure in agricultural activities. At the beginning of agricultural history, seeds were broad cast by hand. The main **defect** of this method is that the seeds could not be sown to a desirable depth and many of them remained on the surface. So, there was a risk for seeds being eaten by birds or being blown away by the wind. Seeds on the surface often failed to come up or came up prematurely, only to be killed by frost or something like that.

Seed plough (耧犁) is a most innovative sowing tool. It was invented during the Warring States Period. The machine consists of a frame, a **hopper** and a tube with a spade that can break the soil. The tube can reach a specific depth under the ground, where seeds are dropped and covered at the same time. At the beginning, it was single-footed and later dual-footed. During Han Dynasty, an officer named **Zhao guo**[2] invented a tri-foot seed plough based on the previous ones, and the new tool can seed in three rows at a time, thus greatly improved the sowing speed.

Seed plough ensures efficient and even growth of crops. With the aid of the tool, farmers got much greater control over the seed planting depth and covered the seeds without back-tracking. Besides, crops planted in this way grow in regular rows, allowing weeding with tools possible.

The technology of multi-tube seed plough making is about 1,100 years ahead of Europe. The tool is still in use today but it has been modified and updated so that seeds can be sowed in many rows simultaneously.

Seed Plough

Irrigating Tools—Shadoof, Windlass and Dragon-Spine Water Lift

With irrigating, agriculture took a giant step forward. Since people came to realize the importance of taking initiative to water farming land, they started to devise many irrigating devices, among which shadoof （桔槔）, windlass （辘轳）, and Dragon-**Spine** Water Lift （龙骨水车） are the most representative ones.

Shadoof was a device used to pick up water from wells in Spring and Autumn Period. It is regarded as one of the earliest examples to put **leverage** theory into practice. In this device, a **beam** supported by a tree or a **column** was divided into two unequal parts. A **weight** was fixed at the end of the shorter part of the beam while a vessel was hung by a rope at the other end. The design of shadoof was practical and convenient. With the help of the weight, the operator could save energy when lifting and lowering the vessel by

Shadoof

pulling up and down the beam and the weight. This is because the moment arm (力臂) on his side is longer than the one on the other side. In this operation, shadoof acted exactly as a lever. It is still not known how shadoof was invented but it is reasonable to assume that the inventor had extensive technical knowledge of the lever. Now, it is still used in some remote Chinese countryside.

Windlass is a kind of water-lifting facility made up of a **pulley**, a crank, a rope, a bucket, an axle and a bracket. Firstly the bracket is fixed over the well and the horizontal axel goes through in the middle of the bracket. The pulley is equipped on the axle to wound ropes. One end of the rope is hitched with a bucket to hang in the well to contain water, and the other end is the fixed handle. In watering, one needs to rotate the handle to pull up the bucket. Windlass is a kind of machinery based on the axle principle.

Windlass

The earliest windlass was created during Qin and Han Dynasties. Actually, at that time no **crank** was installed into the device, so it could only change the direction of the force through the pulley without saving much labor. In Yuan Dynasty, people made some improvements on the device. As an important part of windlass, crank was fixed onto the pulley, which made the lever arm much longer and thereby saved much labor. In some important agricultural books windlass was illustrated with crank, such as *Wangzhen Nongshu* and *Tiangong Kaiwu* written in Yuan and Ming Dynasties. Windlass is still in use today in some countryside.

Dragon-Spine Water Lift is generally listed as one of the most successful and innovative farming tools in ancient China. It was invented by **Bi Lan**[3] in Eastern Han Dynasty and it was perfected and popularized by **Ma Jun**[4] in Three Kingdoms Period, aiming to irrigate farmland in a more efficient way.

The water lift in its early stage was man-powered and comprised of a system of **cogs**, paddles and a chain of wooden pallets—operating similarly to the chain on a bicycle. The movements of pallets helped fetch water from rivers or lakes to farmland. With the advanced mechanical parts, much labor was saved and more water was **pumped** than previous irrigating tools. Since the movements of pallets resemble that of dragons' spine, the tool was named as Dragon-Spine Water Lift.

Dragon-Spine Water Lift

The tool was of high technical level in that period. Cogs and chains in the tool were invented based on mechanical theory and laid a foundation for later development of mechanical equipment. In Southern Song Dynasty, Dragon-Spine Water Lifts were powered by water, and meanwhile animals and wind were also put into use. It was not until the 20th century that the tool was replaced by mechanical **pumps**.

The invention of the above farming tools reflects the originality and creativity of Chinese people. Farming tools liberated people from labor-consuming jobs and increased the productivity. Without them, the agriculture of ancient China couldn't have reached its prosperity.

New words ...

plough [plaʊ] *n.* A plough is a large farm tool with sharp blades that is pulled across the soil to turn it over, usually before seeds are planted. 犁

topsoil [ˈtɒpsɔɪl] *n.* Topsoil is the layer of soil nearest the surface of the ground. 顶层土

durability [ˌdjʊərəˈbɪləti] *n.* 耐用性

moldboard [ˈməʊldbɔːd] the curved blade of a plough, which turns over the furrow 犁壁

plowshare [ˈplaʊʃeə] *n.* The ploughshare is the horizontal pointed cutting blade of a plough 犁铧头

shaft [ʃɑːft] *n.* 辕杆

intermittent [ɪntəˈmɪt(ə)nt] *n.* Something that is intermittent happens occasionally rather than continuously. 断断续续的

defect [ˈdiːfekt] *n.* A defect is a fault or imperfection in a person or thing. 缺陷

hopper [ˈhɒpə] *n.* 送料斗；漏斗

spine [spaɪn] *n.* Your spine is the row of bones down your back. 脊柱

leverage [ˈliːv(ə)rɪdʒ] *n.* 杠杆作用

beam [biːm] *n.* A beam is a long thick bar of wood, metal, or concrete, especially one used to support the roof of a building. 梁

column [ˈkɒləm] *n.* A column is something that has a tall, narrow shape. 柱状物

weight [weɪt] *n.* You can refer to a heavy object as a weight, especially when you have to lift it. （尤其指必须举起的）重物

pulley [ˈpʊlɪ] *n.* A pulley is a device consisting of a wheel over which a rope or chain is pulled in order to lift heavy objects. 滑轮

crank [kræŋk] *n.* A crank is a device that you turn in order to make something move. 曲柄

cog [kɒg] *n.* A cog is a wheel with square or triangular teeth around the edge, which is used in a machine to turn another wheel or part. 齿轮

pump [pʌmp] *v.* To pump a liquid or gas in a particular direction means to force it to flow in that direction using a pump. 抽送
n. a machine for forcing liquid or gas into or out of something 泵；唧筒；抽水机；打气筒

Phrases and expressions

evolve into　发展成，进化成
in contrast to　与…形成对照

come up 发芽
consist of 由…组成
with the aid of 在…的帮助下
take initiative 积极主动
be comprised of 由…组成

Cultural Notes

1. **New Stone Age** begins with the introduction of farming, dating variously from about 9,000 BC in the Near East, 7,000 BC in Southeast Europe, 6,000 BC in East Asia, and even later in other regions. This is the time when cereal cultivation and animal domestication was introduced.

2. **Zhao guo**（赵过）, an agronomist in Western Han Dynasty when Emperor of Wudi was in reign, who made great contributions in innovating agricultural technology and farming tools. The most frequently cited examples include Daitian method, coupling plough method, and Seed Plough according to *Hanshu* edited by Ban Gu.

3. **Bi Lan**（毕岚）, was one member of a eunuch gang in the Eastern Han Dynasty which was infamous for its manipulating of the political power and bringing disasters and chaos to the country. In spite of that, he was also remembered as the inventor of the Dragon-Spine Water Lift which was originally invented to draw water from the river to sprinkle roads.

4. **Ma Jun**（马钧）, living in Wei in the Three Kingdoms Period, was one of the most famous mechanical inventors in the history of science and technology in China. He improved the water lift invented by Bi Lan and put it to use in irrigating farming land.

Exercises

1. Read the following statements and try to decide whether it is true or false according to your understanding.

_____ 1) The earliest tools used for soil preparation in China was *lei* and *si*.

_____ 2) The curved-shaft plough was a representative achievement of the renovation of agricultural machinery in Han Dynasty.

_____ 3) Seed Plough was invented earlier than plough.

_____ 4) In Yuan Dynasty, people made some improvements on windlass by fixing a crank onto the pulley.

_____ 5) The high working performance of Dragon-Spine Water Lift made it widely used throughout Asia.

2. Fill in the blanks with the information you learn from the text.

1) _____ is the breaking through invention in preparing soils.

2) Moldboard plough composed of a triangular-shaped _____ and a U-shaped _____.

3) The most important benefit of curved-shaft plough lies in that it can adjust _____ flexibly.

4) With the aid of seed plough, farmers got much greater control over the seed planting _____. Besides, crops planted in this way grow in regular _____, allowing weeding with tools possible.

5) _____ is considered as one of the earliest examples that applying leverage theory into practice.

3. Explain the following terms.

1) Plough

2) Seed Plough

4. Translate the following paragraph into English.

犁的发明具有重要历史意义，因为它是连续运动的农具，而之前的农具如耒、耜、铲、锄等都是断续运动的农具，劳动生产率因之大大提高，这就为犁耕农业的产生创造了必要的技术条件，可谓原始农业上的巨大进步。

5. Critical thinking and discussion.

Modern agricultural machinery in China is developed based on ancient agricultural tools and is being developed at a very fast pace. It is undeniable that technological innovation has played an important role in this process. So how do you think universities should cultivate college students' innovation awareness and creative ability in order to contribute to the development of science and technology in China in the future?

Chapter 5
Flood Control and Irrigation

Section A Yu Taming the Flood

   In the initial stage of human civilization, floods happened frequently and brought about disasters to the world. According to *the Bible*, human beings became helpless and desperate, and almost all were drowned. Only Noah and his family had survived with the aid of the ark and continued to grow by God's arrangement. In reality, ancient Chinese people spared no efforts to protect on their own from the floods. The representative figure in water control is Yu (禹), an emperor of the Xia Dynasty whose work of water conservancy brought about agricultural growth and a more safe, stable and peaceful society. In addition, a substantial number of water conservancy facilities emerged, including those used in flood control, irrigation, shipping, aquaculture and so on. Among them the Dujiangyan Irrigation System was very influential. The stories of people like Yu or water projects about flood control reflect ancient people's wisdom and philosophy of life.

   Nature has not simply brought humans opportunities but also some threats. All around the world people have been telling stories about great floods, and it seems that floods about 4,000 years ago caused devastation to many of the first civilizations, including China. But China managed to survive the disasters through efforts and wisdom. One of such stories is *Yu the Great Harnessing the Waters*, called *da yu zhi shui* (大禹治水) in Chinese, in which *zhi shui* means controlling or taming flood.

   According to the journal *Science* there were nine-year-long great floods around the 20th century BC from heavy rains and the outbreak of some **barrier lakes**[1] in modern Qinghai Province. They caused the Yellow River to change its course in many places with devastating effects.

   In addition to natural factors, social reasons were also responsible for the floods. The destruction of the forests and grassland from the primitive agriculture greatly led to the soil and water erosion, and as a result, the Yellow River often flooded a vast area in ancient China.

For thousands of years the Yellow River regularly burst its banks, wiping out entire villages in its path. As ***Records of the Grand Historian***[2] said, "it filled the empty valleys, flooded the flat land, and even surrounded the mountains and hills." Many People were killed, cattle were drowned, some crops and houses were destroyed or damaged, so it is not difficult to know why the Yellow River was known as China's Sorrow.

In order to get flood victims out of trouble, different people were involved in the work of flood control. A case in point is Yu the Great, whose story has been inspiring Chinese people to overcome obstacles and pursue happiness.

The old legends say that one of the tribe leaders Yao（尧）appointed a man named Gun（鲧）to devise a way to tame the river. Gun, the father of Yu, adopted a method of building thick walls with earth to stop water flooding onto very low-lying land. However, it no longer worked well nine years later. Powerful flood water broke the dike and brought disaster to the people again. And Gun was severely punished for failing to accomplish the task.

The father's burden would now fall upon his son Yu（禹）after Gun's execution. Another tribe leader Shun（舜）ordered Yu to come up with a new idea about how to control the floods, and Yu dedicated his life to the job. According to old legends, Yu said he wouldn't return to his wife who was then pregnant until the river was tamed.

Yu Taming Water

Legend has it that Yu decided to begin by surveying the entire length of the river. On this long trip on foot he figured out a radically different plan, which was no more **confrontations** with nature and no more dams. Instead of trying to obstruct the raging waters like his father, he would divide them. Then Yu created a vast network of channels, which meant a large amount of work of engineering. During the flood season, the channels would **divert** the rushing river and reduced the risk of destruction. Finally, his vast network of channels was completed and allowed the flood waters to flow along river courses into the sea. And when the rains came again, his great feat of engineering stood the test. Over the 13 years, Yu passed his home three times, but he never went inside, which showed his true **self-sacrifice**.

In the struggle to harness the floods, Yu tried his best to successfully convince hundreds of rival tribes to set aside years of **hostility**. And he also united the tribes of the Yellow River for the first time. Under his leadership, the tribes eventually found strategies to bring the serious Yellow River flooding under control. As a reward, the leader Shun made Yu his **heir**. Thus it's not hard to understand that some believe Yu founded the first Chinese dynasty, Xia Dynasty. And it's commonly accepted that Chinese civilization began on the banks of the Yellow River. Yu's heroic deeds were also recorded in some ancient classics. ***The Book of Songs***[3] recorded, "the flood was boundless, and Yu compressed the earth." *Records of the Grand Historian* also said, "Dayu managed water and soil, proving himself a man as great as heaven".

There is no doubt that Yu and his efforts played an important role in the development of Chinese civilization in ancient times. The ultimate triumph of his **ingenuity** brought order to the land so that fields could be cultivated. People in this region also improved their farming methods and had access to the Yellow River water for irrigation. Since then, China has begun in a true sense to have agriculture, which has **fueled** the blossoming of Chinese civilization. With the development of agriculture and fusion of tribes, the Xia Dynasty emerged in Chinese history.

It is not difficult to find that Yu's water control principle "dredge and channel rivers to drain the flood waters" has had far-reaching impacts on those who harness rivers later. Although human efforts are important in conquering nature, it's more important for people to respect laws of nature and then take actions **in light of** changing circumstances. And this concept or value has been gradually adopted to address problems in actual reality and incorporated into Chinese culture.

New words

confrontation [ˌkɒnfrʌnˈteɪʃn] *n.* A confrontation is a dispute, fight, or battle between two groups of people. 对抗冲突；战斗；战役

divert [daɪˈvɜːt] *v.* To divert vehicles or travelers means to make them follow a different route or go to a different destination than they originally intended. （使）转向；（使）绕道

self-sacrifice [ˈselfˈsækrəfaɪs] *n.* Self-sacrifice is the giving up of what you want so that other people can have what they need or want. 自我牺牲

hostility [hɒˈstɪləti] *n.* Hostility is unfriendly or aggressive behaviour towards people or ideas. 敌意；对抗；敌对行为

heir [eə(r)] *n.* An heir is someone who has the right to inherit a person's money, property, or title when that person dies. 继承人

ingenuity [ˌɪndʒəˈnjuːəti] *n.* Ingenuity is skill at working out how to achieve things or skill at inventing new things. 心灵手巧；善于创造发明

fuel [ˈfjuːəl] *v.* To fuel a situation means to make it become worse or more intense. 刺激；使变得更糟；使加剧

Phrases and expressions

in light of   If you do or decide something in light of a new situation or new information, you do it because of that situation or information 按照，依据

An Overview of Chinese Farming Culture

Cultural Notes ...

1. barrier lake (also earthquake Lake) is a lake formed after an earthquake strikes. Around 1920 BC, the earthquake created a massive landslide tons of rock, which stopped the flow of the Yellow River in its water course.

2. *Records of the Grand Historian* (《史记》) is called *Shiji* in Chinese. It is a monumental history of ancient China and the world finished around 94 BC by the Han Dynasty official Sima Qian after having been started by his father, Sima Tan, Grand Astrologer to the imperial court. The work covers a 2,500-year period from the age of the legendary Emperor Yellow to the reign of Emperor Wu of Han. The *Records* set the model for editing the 24 subsequent dynastic histories of China, breaking it up into smaller, overlapping units dealing with famous leaders, individuals, and major topics of significance.

3. *The Book of Songs* (《诗经》) is the oldest existing collection of Chinese poetry, comprising 305 works dating from the 11th to 7th centuries BC. It is said to have been compiled by Confucius, and has been studied and memorized by scholars in China and neighboring countries over two millennia. Since the Qing Dynasty, its rhyme patterns have also been analyzed in the study of old Chinese phonology.

Exercises ...

1. Read the following statements and try to decide whether it is true or false according to your understanding.

_____ 1) In history landslides during the earthquake once blocked the Yellow River, creating lakes and posing risks of flooding communities downstream.

_____ 2) Gun built huge dams with stones and earth across the Yellow River, which prevented the floodwaters from damaging crops and houses.

_____ 3) Instead of trying to control the raging waters like his father, Yu

Chapter 5 Flood Control and Irrigation

would divide them and reduce the risk of floods.

____ 4) It was the legendary Yellow Emperor that united the clans and tribes of the Yellow River Valley for the first time.

____ 5) The story of *Da yu zhi shui* tells us with efforts man can conquer nature.

2. Fill in the blanks with the information you learn from the text.

1) According to *Science* there were nine-year-long great floods from heavy rains and the outbreak of some _____ in modern Qinghai Province.

2) Gun, the father of Yu, adopted a method of _____ to stop water flooding onto very low-lying land.

3) Another tribe leader _____ ordered Yu to come up with a new idea about how to control the floods, and Yu dedicated his life to the job.

4) Some people think Yu founded the first Chinese dynasty—_____, and it is well accepted that Chinese civilization began on the banks of the Yellow River.

5) Although human efforts are important in dealing with nature, it's more important for people to _____ and then take actions in accordance with _____ .

3. Explain the following terms.

1) *Da yu zhi shui*
2) Yu's Flood Control Principle

4. Translate the following paragraph into English.

黄河（the Yellow River）是中华民族的摇篮。她孕育了一代又一代的中国人，但她引发的洪水也给人们带来了灾难。勤劳的中国人民世代都在努力治理黄河水。大禹是4 000多年前中国著名的治水家。大禹吸取了他父亲曾经治水的经验教训，经过大量的调查和研究，发现了引发洪水的原因。大禹工作很努力，他致力于治水13年。在这段时间里，他没有回过家，甚至三过家门而不入。

5. Critical thinking and discussion.

As is known to all, Yu the Great couldn't have managed the floodwaters had it not been for his creative way. How was his way different from his father's? And what's the inspiration behind the story of Yu controlling floodwaters?

Section B Dujiangyan Irrigating System

Floods are frequent and devastating events worldwide. China is one of the flood-stricken countries. As floods have commonly occurred, efforts have been made to raise awareness of flood risks and flood prevention. To avoid death tolls and economic loss caused by **devastating** flood disasters, effective measures have been adopted. Some 4,000 years ago to address the issue, the legendary figure Yu successfully dredged channels and diverted water of the Yellow River.

Later emperors and officials also attached much importance to flood control. During the Warring States Period, **Guan Zhong**[1], the prime minister of Qi state, once said, "A good emperor must first eliminate five evils including floods, droughts, storms, epidemics and plagues of insects, then he could manage his people successfully. And flood disaster is taking the first place." He proposed strategic measures in managing rivers and encouraged the king Qi Huangong（齐桓公）to pay attention to flood control and irrigation. Another leader paying close attention to water conservancy is the King Zhaoxiang of Qin State（秦昭襄王）, the great-grandfather of **Qin Shihuang**[2], the emperor of

Li Bing Heading the Dujiangyan Irrigation Project

Qin state. He believed that the development of agriculture could lead the country to prosperity and power. And he thought that the water supply and flood control were of primary importance for developing agriculture. So the hydrologist **Li Bing**[3] was assigned to lead thousands of people to design and construct the Dujiangyan Irrigation System in 256 BC.

Origin of the Dujiangyan System

The Dujiangyan system is one good case of ancient irrigation and flood control projects in China. As an irrigational **infrastructure**, it is located in Dujiangyan City of Sichuan Province, near the capital Chengdu City. It is still in use today to irrigate over 5,300 square kilometres of land in the region. The Dujiangyan, the Zhengguo Canal in modern Shaanxi（陕西）Province and the Lingqu Canal in modern Guangxi Zhuang Autonomous Region are collectively known as the "three great **hydraulic** engineering projects of the Qin state."

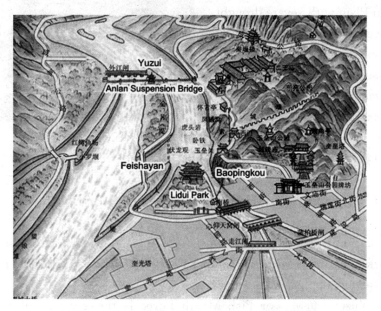

Bird's-eye View of the Dujiangyan Irrigation System

The system's infrastructure is on the **Min River**, the longest **tributary** of the Yangtze River. It is in the west part of the Chengdu Plain, and at the confluence between the Sichuan basin and the Tibetan plateau. Originally the

Min River rushed down from the Min Mountains, but slowed abruptly after reaching the Chengdu Plain, filling the watercourse with sand, which made the nearby areas extremely prone to floods.

During the Warring States Period, people who lived along the banks of the Min River were plagued by annual flooding. Li Bing, an irrigation engineer, investigated the problem and discovered that the river was swelled by fast flowing spring melting water from the local mountains. It burst the banks when reaching the slow moving and heavily silted stretch below. Li Bing, the governor of Shu (蜀) state, and his son headed the construction of the Dujiangyan. He proposed to construct an artificial dam to redirect a portion of the river's flow. Li Bing received 100,000 taels of silver for the project from the government and set to work with a team of thousands of workers. And then he guided people to cut a channel through Mount Yulei (玉垒山) to discharge the excess water upon the Chengdu Plain with severe drought.

Macha

Together with the channel, Li Bing also constructed a levee with long sausage-shaped baskets of woven bamboo filled with stones, known as *Zhulong* (竹笼) held in place by wooden tripods known as *Macha* (杩槎). In end, this engineering took four years to finish.

Main Components of Dujiangyan Irrigation Project

Li Bing's Irrigation System consists of the following main constructions that work in harmony with one another to effectively control flooding and keep the fields well supplied with water.

The Yuzui (鱼嘴) or Fish Mouth Levee is the key part of Li Bing's construction, named for its cone-shaped head that resembles the mouth of a fish. It is an artificial **levee** that divides the water into inner and outer streams. The inner stream bed is deep and narrow, while the outer stream bed is relatively shallow but wide. This special structure ensures that the inner

Chapter 5 Flood Control and Irrigation

Fish Mouth Levee

stream carries approximately 60% of the river's flow into the irrigation system during dry season. While during flood, this amount decreases to 20% to protect the people from flooding. The outer stream drains away the rest, flushing out much of the **silt** and **sediment.**

The Feishayan (飞沙堰) or Flying Sand **Weir** has a 200-meter-wide opening that connects the inner and outer streams. It provides protection against flooding by allowing the natural **swirling** flow of the water to drain out excess water from the inner to the outer stream. The swirl also drains out silt and sediment that fail to go into the outer stream. A modern reinforced concrete weir has replaced Li Bing's original weighted bamboo baskets.

The Baopingkou (宝瓶口) or Bottle-Neck Channel, which Li Bing **gouged** through the Yulei Mountain, is the final part of the system. The channel distributes the water to the farmlands in the Chengdu Plain. As a check gate, the narrow entrance creates the whirlpool flow that carries away the excess water from Flying Sand Weir, and thus this can effectively prevent flooding. Interestingly, the name Baopingkou derives from the narrow entrance.

The channel between Mount Yulei and **Li Pile** was **excavated** by Li Bing to let the water through. The swirling flow happens to form due to water in the inner stream hitting against the rocks of Li Pile. Over 2,000 years ago in the State of Qin with no gunpowder or ironware used widely, it was incredibly difficult to excavate the Yulei Mountain. But Li Bing and the workers cut the rock layers by burning the rocks with firewood, then pouring water or vinegar on them. After repeating the process again and again over eight years, they finally made a gap 20 meters wide, 40 meters high, and 80 meters long. Eventually the gap separated Li Pile from Mountain Yulei and then the Baopingkou came into being.

Bottle-neck Channel

After the Dujiangyan System was finished, no more floods occurred. And people in Chengdu Plain could cultivate on the irrigated farmland. The area developed quickly in agriculture and provided solid economic foundation for the establishment of Qin Dynasty, the first unified feudal regime of China. For over 2,200 years, Dujiangyan Irrigation System has enabled Sichuan to have regular and bountiful harvests and earn the name *tian fu zhi guo*（天府之国）in Chinese, meaning "a country of heaven" or "a land of abundance". That's why people built a **shrine** in honor of Li Bing on the east side of Dujiangyan.

Today, Dujiangyan has become a major tourist attraction. It is also admired by experts and engineers from around the world, because unlike contemporary dams where the water is blocked with a huge wall, Dujiangyan still lets water go through naturally. Modern dams do not let fish go through very well, since each dam is a wall and the water levels are different. In 2000, Dujiangyan became a **UNESCO World Heritage Site**.

Chapter 5　Flood Control and Irrigation

New words ...

devastating [ˈdevəsteɪtɪŋ] *adj*. If you describe something as devastating, you are emphasizing that it is very harmful or damaging. 毁灭性的；极具破坏性的

infrastructure [ˈɪnfrəstrʌktʃə(r)] *n*. The infrastructure of a country, society, or organization consists of the basic facilities such as transport, communications, power supplies, and buildings, which enable it to function. （国家、社会、组织赖以行使职能的）基础建设，基础设施

hydraulic [haɪˈdrɔːlɪk] *adj*. Hydraulic equipment or machinery involves or is operated by a fluid that is under pressure, such as water or oil. （设备或机械）水压的，液压的

tributary [ˈtrɪbjətəri] *n*. A tributary is a stream or river that flows into a larger one. 支流

levee [ˈlevi] *n*. A low wall built at the side of a river to prevent it from flooding 防洪堤

silt [sɪlt] *n*. Silt is fine sand, soil, or mud which is carried along by a river. 淤泥；泥沙

sediment [ˈsedəmənt] *n*. Sediment is solid material that settles at the bottom of a liquid, especially earth and pieces of rock that have been carried along and then left somewhere by water, ice, or wind. 沉渣；沉淀物

weir [wɪə] *n*. A weir is a low barrier which is built across a river in order to control or direct the flow of water. 堰；拦河坝（使））扩大；（使）增加

swirl [swɜːl] *v*. If you swirl something liquid or flowing, or if it swirls, it moves round and round quickly （使）打旋；（使）旋动；（使）起旋涡

gouge [ɡaʊdʒ] *v*. If you gouge something, you make a hole or a long cut in it, usually with a pointed object. 凿；挖

excavate [ˈekskəveɪt] *v*. When archaeologists or other people excavate a piece of land, they remove earth carefully from it and look for things such as pots, bones, or buildings buried there, in order to discover information about the past. 发掘，挖掘（古物等）

shrine [ʃraɪn] n. A shrine is a place of worship which is associated with a particular holy person or object. 神庙；神龛

Phrases and expressions

raise awareness of 增强意识
be prone to 有…的倾向，易于

Proper names

Min River 岷江，长江上游的一条重要支流，位于四川省境内。
Li Pile 离堆，位于玉垒山，李冰为建都江堰工程，开凿玉垒山为两部分，小而低的那部分山体被称为"离堆"。
UNESCO World Heritage Site 联合国教科文组织世界遗产

Cultural Notes

1. **Guan Zhong** (725 BC-645 BC) was a chancellor and reformer of the State of Qi during the Spring and Autumn period of Chinese history. He is mainly remembered for his reforms as chancellor under Duke Huan of Qi, as well as his friendship with his colleague Bao Shuya, though his reputation remained controversial among the Confucians of his time.

2. Qin Shihuang(259 BC-210 BC) was the founder of the Qin Dynasty and was the first emperor of a unified China. He became the King Zheng of Qin(秦王政)when he was thirteen, then China's first emperor when he was 38 after the Qin had conquered all of the other Warring States and unified all of China in 221 BC. Rather than maintain the title of "king" borne by the previous Shang and Zhou rulers, he ruled as the First Emperor(始皇帝)of the Qin dynasty from 220 to 210 BC. His self-invented title "emperor" would continue to be borne by Chinese rulers for the next two millennia.

3. Li Bing (about 302 BC-235 BC) was a Chinese irrigation engineer and politician of the Warring States period. He served the state of Qin as an administrator and has become renowned for his association with the construction of the Dujiangyan Irrigation System, which he is traditionally said to have prompted and overseen. Because of the importance of this 2000-year-old irrigation system to the development of Sichuan and the Yangtze River region, Li Bing became a great Chinese cultural icon, hailed as a great civil administrator and water conservation expert.

Exercises

1. Read the following statements and try to decide whether it is true or false according to your understanding.

_____ 1) Dujiangyan is an irrigation system built in the Spring and Autumn Period of China by the Qin State.

_____ 2) Guan Zhong once said a sensible emperor should give priority to control of floods, droughts, storms, epidemics and plagues of insects.

_____ 3) Statistically, the outer stream carries approximately 60% of the river's flow during dry season.

_____ 4) Baopingkou is a key link of the irrigation system, working as a strategic check gate controlling the water to the farmlands in the Chengdu Plain.

_____ 5) Without powder or ironware, the ancient people still managed to excavate the Yulei Mountain with its hard sedimentary rock.

2. Fill in the blanks with the information you learn from the text.

1) The King Zhaoxiang of Qin State assigned the hydrologist Li Bing in 256 BC to lead thousands of people to design and construct _____ .

2) The Dujiangyan, _____ and _____ are collectively known as the "three great hydraulic engineering projects of the Qin state".

3) _____, the key part of Li Bing's construction, is an artificial levee that divides _____ into inner and outer streams.

4) After eight years of efforts, Li Bing and his followers made a gap separating Li Pile from _____ and then _____ came into being.

5) For over 2,200 years, Dujiangyan Irrigation System has enabled Sichuan to have regular and bountiful harvests and earn the name _____ in Chinese, meaning "a country of heaven" or "_____".

3. Explain the following terms.

1) Fish Mouth Levee
2) the Bottle-Neck Channel

4. Translate the following paragraph into English.

都江堰水利工程是由蜀郡守李冰于公元前256年左右率众修建的，距今已有2 200多年的历史，是全世界迄今为止年代最久、唯一留存、以无坝引水为特征的宏大水利工程。成都平原形似一把张开的纸扇向东南倾斜，而都江堰市恰好处在扇形平原顶端，海拔700多米。李冰合理利用了这种自然倾斜的地形，在不破坏自然资源情况下修建了都江堰。都江堰渠首枢纽主要由鱼嘴、飞沙堰和宝瓶口三部分组成。都江堰建成后，成都平原沃野千里，使成都有了"天府之国"的美誉。都江堰不仅是闻名于世的古代水利工程，也是著名的风景名胜区。2000年都江堰被联合国教科文组织作为文化遗产列入《世界遗产名录》。

5. Critical thinking and discussion.

Dujiangyan Irrigation System has been playing a major role in flood control and irrigation since the completion 2,000 years ago. Try thinking about why Li Bing didn't straighten the watercouse but built it in the shape of curves, and how the great man-made system would inspire modern water conservancy projects?

Section C Harnessing of the Yellow River

The Yellow River is known in Chinese as Huang He, and it is called the "Mother River of China" and "the Cradle of Chinese Civilization." But it is also called "China's Sorrow" because it is extremely prone to flooding.

In the past, each year tons of soil flowed into the Yellow River, and this causes the continual rise and shift of the riverbed. Besides, it had a large flow rate, because it flowed from east of Tibetan Plateau to west of North China Plain with a great difference in altitude. In addition, the frequent rainfalls caused poor **drainage** in its lower reaches due to the climate there. What's more, its course often changed and the river mouth sometimes changed catastrophically hundreds of kilometers. It had about 26 major changes in course in the past 2,000 years. So the floods often occurred in the Yellow River Basin in ancient times. When floods occurred, some of the population might initially die from drowning and then many more would suffer from the **ensuing** famine and spread of diseases. Also the surrounding farm fields would be damaged seriously. So Huang He flooding control has been a big concern of both ancient and present Chinese people.

Throughout history considerable efforts have been made to strengthen its levees and control its floods. Before Qin state united China, **Zheng Guo**, a water engineer helped the state of Qin to build a canal along the upstream of the Yellow River in 246 BC and it irrigated the salty land of the Guanzhong Plain, north of Xi'an. This canal was called Zhengguo Canal, which was one of the early flood control projects in the Yellow River Basin. After the area was thus turned into fertile land, Qin became rich and strong, and in the end unified the **feudal** states.

The Han Dynasty that followed the Qin continued to promote large flood control, irrigation, and navigation projects on the Yellow River. At that time, frequent earthquakes in the area not only killed thousands of people directly, but blocked the river with **landslide debris** that **catastrophically** flooded downstream areas. So major human efforts would be required to build and maintain levees and canals to control floods, with irrigation and navigation perceived as additional benefits.

In Western Han Dynasty an engineer **Jia Rang** proposed three measures to manage floods. The best policy was to channelize the river, improving its rate of flow to the sea but at the cost of fields and towns. The second best plan was to divert the excess water so that floods would be **mitigated**. And the unwise decision was to build higher levees every year. Jia Rang's strategy was recorded in the ancient geography book **Hanzhi** and had a profound influence on the later work of river management.

An Eastern Han Dynasty successor **Wang Jing** stabilized the Yellow River with levees in a way that lasted for centuries. Wang Jing personally surveyed the **terrain** and planned the **embankment**. First, he strengthened the dikes of the Yellow River, built many diversion channels and constructed water gates. And then he dredged and straightened **Bianqu Canal** linking the Yellow River and the Huai He. Wang Jing's work achieved great success. As a result, in the following 800 years after Wang Jing the Yellow river and the connecting rivers were calmed and very few flood disasters occurred.

Then in Ming Dynasty, Pan Jixun, a water conservancy expert, **presided** over governing the Yellow River for four times, which lasted for 27 years. In the long-term practice, he learned from the **predecessor**s and summed up Chinese people's experience in water control. Consequently, he regulated the levee system again, blocked many branches of the river and made the river flow in a main channel. What's more, he built two layers of **dikes** to confine floods. Most importantly, he proposed the strategy of flushing sediment with **converging** flows by narrowing the channel, which had a great impact on later generations to deal with the river control. His philosophy of flood prevention and river regulation had significantly changed from diversion to embankment ever since.

Over the ages past dynasties exerted great efforts to prevent floods in order to establish a strong country where their people could enjoy a wealthy life, because the frequent river floods were a barrier to agricultural growth. In the Qing Dynasty, the Yellow River floods still influenced national economy and people's livelihood. So many talented administrators of river engineering had done much in river regulation. Among them, **Jin Fu**

Chapter 5 Flood Control and Irrigation

and **Chen Huang** enjoyed a higher reputation. Jin and Chen applied Pan's theory and practiced "converging flow with narrow channel and **scouring** sand with high flow rate". Both of them made contributions to advancing flood control techniques.

Pan Jixun, Jin Fu and Chen Huang tried to control floods by constructing the artificial dams to regulate the water flow. Their measures embody **Confucianism**[1], which is of discipline and order imposed upon nature. It contrasts with the **Taoism**[2] solution of allowing the river a more "natural" course within lighter **constraints**. The flooding projects before Ming Dynasty generally reflected Taoism of river management. In other words, the former river management represented a tremendous interference with nature, and the latter showed more respects for it.

The battle of managing the Yellow River has been far from over. Nowadays Chinese government has built a series of water projects to control floods. Among them, **Sanmenxia Hydropower Station** has been playing an important role in regulating floods and reducing droughts. More importantly, China also advocates **reafforestation** and soil conservation to cut down the supply of **eroded** sand to Huang He. For example, Three-North Shelter Forest Program aims to build a network of forests in all three northern regions including the North, the Northeast and the Northwest from 1979 to 2050. While this project is not yet to cure all the problems it will have a positive impact. The program has begun to play a part in fixing water and soil and improving the ecological environment of Loess Plateau. Also, floods are being greatly controlled and farmland is effectively irrigated in the Yellow River basins.

Now The Yellow River Basin grows more than 50% of China's wheat, corn, and cotton etc. The irrigation projects built on the river have helped increase the crop **yield**s to feed Chinese people, one of the densest populations in the world, with **surplus** usually to feed people in other parts of the world. With its important position in China, the issue of managing and protecting the Yellow River cannot be stressed too much anytime in the future.

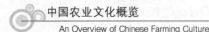

drainage [ˈdreɪnɪdʒ] *n.* Drainage is the system or process by which water or other liquids are drained from a place. 排水系统；排水；放水

ensuing [ɪnˈsjuːɪŋ] *adj.* Ensuing events happen immediately after other events. 接着发生的；接踵而来的

feudal [ˈfjuːdl] *adj.* Feudal means relating to the system or the time of feudalism. 封建体制的；封建时期的

landslide [ˈlændslaɪd] *n.* A landslide is a large amount of earth and rocks falling down a cliff or the side of a mountain. 塌方；山崩；滑坡

debris [ˈdebriː] *n.* Debris is pieces from something that has been destroyed or pieces of rubbish or unwanted material that are spread around. （被毁物的）残骸；碎片；垃圾

catastrophic [ˌkætəˈstrɒfɪk] *adj.* Something that is catastrophic involves or causes a sudden terrible disaster. 灾难性的；引起重大灾难的

mitigate [ˈmɪtɪgeɪt] *v.* To mitigate something means to make it less unpleasant, serious, or painful. 减轻；缓解

terrain [teˈreɪn] *n.* Terrain is used to refer to an area of land or a type of land when you are considering its physical features. 地形；地势；地带

embankment [ɪmˈbæŋkmənt] *n.* An embankment is a thick wall of earth that is built to carry a road or railway over an area of low ground, or to prevent water from a river or the sea from flooding the area. （公路、铁路等的）路堤；（河、海等的）堤，堤岸

preside [prɪˈzaɪd] *v.* If you preside over a meeting or an event, you are in charge. 负责；主持

predecessor [ˈpriːdəsesə] *n.* Your predecessor is the person who had your job before you. 前任；前辈

dike [daɪk] *n.* A long thick wall that is built to stop water flooding onto a low area of land, especially from the sea 堰；堤；坝

converge [kənˈvɜːdʒ] *v.* If people or vehicles converge on a place, they move towards it from different directions. 汇集；集中；聚集

scour [skaʊə] *v.* If you scour something such as a sink, floor, or pan, you clean its surface by rubbing it hard with something rough. 擦净；擦亮

constraint [kənˈstreɪnt] *n.* A constraint is something that limits or controls what you can do. 限制；束缚；约束

reafforestation [ˌriːəˌfɒrɪˈsteɪʃn] *n.* (＝) Reforestation of an area where there used to be a forest is planting trees over it. 重新造林；再植树；重新绿化

erode [ɪˈrəʊd] *v.* If rock or soil erodes or is eroded by the weather, sea, or wind, it cracks and breaks so that it is gradually destroyed. （使）侵蚀；（使）腐蚀；（使）风化

yield [jiːld] *n.* A yield is the amount of food produced on an area of land or by a number of animals. （农作物或牲畜的）产量，产出

surplus [ˈsɜːpləs] *adj.* Surplus is used to describe something that is extra or that is more than is needed. 过剩的；剩余的；多余的

Phrases and expressions

be credited with 被认为，归功于

impose upon If you impose something on people, you use your authority to force them to accept it. 强制实行强加

interfere with Something that interferes with a situation, activity, or process has a damaging effect on it. 妨碍；冲突；抵触

Proper names

Zheng Guo 郑国，战国时期韩国卓越的水利专家，所修郑国渠和都江堰、灵渠并称为秦代三大水利工程。

Jia Rang 贾让，中国西汉时期筹划治理黄河的代表人物。生卒年不详。因提出治理黄河的上、中、下三策而著名。

HanZhi 《汉志》，《汉书》十志的简称，多用于指代《汉书·理志》和《汉书·艺文志》，东汉班固撰。

Wang Jing 王景（约公元 30—85）东汉时期著名的水利工程专家。

Bianqu Canal 汴渠，中国古代沟通黄河和淮河的骨干运河。

Jin Fu 靳辅（1633—1692），清朝水利专家。

Chen Huang 陈潢（1637—1668），清朝水利专家。

Sanmenxia Hydropower Station 三门峡水电站是黄河干流兴建的第一座大型水利枢纽工程，被誉为"万里黄河第一坝"。

Cultural Notes

1. **Confucianism** also known as Ruism, is described as tradition, a philosophy, a religion, a humanistic or rationalistic religion, a way of governing, or simply a way of life. Confucianism developed from what was later called the Hundred Schools of Thought from the teachings of the Chinese philosopher Confucius (551-479 BC). Confucianism revolves around the pursuit of the unity of the individual self and Heaven (Tian), or, otherwise said, around the relationship between humanity and Heaven. The principle of Heaven （天理），is the order of the creation and the source of divine authority, monistic （一元论的）in its structure. Individuals may realize their humanity and become one with Heaven through the contemplation of such order. This transformation of the self may be extended to the family and society to create a harmonious fiduciary community.

2. **Taoism** also known as Daoism, is a religious or philosophical tradition of Chinese origin which emphasizes living in harmony with the *Tao* （道）. The *Tao* is a fundamental idea in most Chinese philosophical schools; in Taoism, however, it denotes the principle that is the source,

Chapter 5　Flood Control and Irrigation

> pattern and substance of everything that exists. Taoism differs from Confucianism by not emphasizing rigid rituals and social order. Taoist ethics vary depending on the particular school, but in general tend to emphasize *wu wei* (effortless action): naturalness, simplicity, spontaneity.

Exercises ...

1. Read the following statements and try to decide whether it is true or false according to your understanding.

____ 1) The Zhengguo canal was partly responsible for the fact that the Qin state eventually became the first to unify China, under the Emperor Qin Shi Huang.

____ 2) Pan Jixun regulated the levee system, blocked many branches of the river and made the river flowing in wider channels in the Ming Dynasty.

____ 3) Chinese government has built a series of water projects to control floods. Sanmenxia Hydropower Station has been playing a part in regulating floods and reducing droughts.

____ 4) The Yellow River is home to crops such as wheat, corn and cotton etc.

____ 5) The narrow-channel encourages fast flow that keeps sediment in suspension, and little capacity for absorbing a major flood crest, and even the high levees overflowing.

2. Fill in the blanks with the information you learn from the text.

1) The Yellow River is known as the "Mother River of China" and "_____". But it is also called "_____" for frequent flooding.

2) In Western Han Dynasty an engineer _____ proposed three measures to manage floods which had a profound influence on the later work of river management.

3) In _____, Pan Jixun, a water conservancy expert, presided

over governing the Yellow River for four times. He blocked many branches of the river and made the river flow in _____ . What's more, he built _____ to confine floods.

4) In the Qing Dynasty, of many talented administrators of river engineering, _____ and _____ enjoyed a higher reputation.

5) Chinese government has advocated _____ aiming to build a network of forests in all three northern regions including the North, the Northeast and the Northwest from 1979 to 2050.

3. Explain the following terms.

1) Confucian solution in controlling the Yellow River
2) Taoist solution in controlling the Yellow River

4. Translate the following paragraph into English.

黄河是亚洲第三、世界第六长的河流。"黄"这个字描述的是其河水浑浊的颜色。黄河发源于青海，流经九个省份，最后注入渤海。黄河是中国赖以生存的几条河流之一。黄河流域（river basin）是中国古代文明的诞生地，也是中国早期历史上最繁荣的地区。然而，由于极具破坏力的洪水频发，黄河曾造成多次灾害。在过去几十年里，政府采取了各种措施防止灾害发生。

5. Critical thinking and discussion.

The Yellow River flows across a great blanket of loess and the river erodes the loess easily, carrying along large amount of sediment load until it reaches the lowland of north China, still 1,400 km from the sea. What would you suggest to avoid the soil erosion in the Loess Plateau? And how do you think to protect the ecological system in the Yellow River basin?

Chapter 6
Traditional Festivals and Folk Customs

Section A Traditional Festivals

A lot of traditional Chinese festivals are celebrated each year, most of which take place on important dates in the Chinese lunar calendar. Their formation is a long process of historical and cultural accumulation in the long history of China. Different festivals are celebrated in different places for China is a vast land and has diverse ethnic groups. Even the same festival has different customs. The traditional festivals have absorbed **nourishment** from various ethnic cultures, and thus they possess rich cultural **connotations**. Undoubtedly, they have become an important and brilliant part of Chinese culture and a precious cultural heritage for the whole Chinese nation.

The Spring Festival

The Spring Festival is regarded as the most important festival by Chinese people, which falls on the first day of the first month. It has **evolved** from Start of Spring of the 24 solar terms, which marks the beginning of spring plowing. Thus it is closely related to agricultural activities in spring.

Usually the Spring Festival is celebrated from the 23rd day of the twelfth month to the 15th day of the first month in the coming year. Various customs are related to the Spring Festival. Usually from the 23rd day to the eve of the Spring Festival, people engage themselves in the preparation for the new year. They offer sacrifices to the Kitchen God, clean their houses, greet the **Jade Emperor**[1], stew pork meat, kill roosters, make steamed

Firecracker Playing

breads, paste red couplets and hang red lanterns, etc. On the eve of Spring Festival, people stay up late into the night to wait for the arrival of the new year. At 12 o'clock in the midnight, people eat *jiaozi* in some places to welcome the new year. And people play firecrackers to drive away **demons**, whose merry sounds also bring joyous atmosphere to the festival.

During the festival, people greet and visit their friends and relatives. In some places, the juniors kowtow to their seniors and when doing so young children will receive *hongbao*, meaning lucky money. Currently giving a New Year greeting on the phone becomes more and more popular. On the 15th day of the first month, people celebrate the Lantern Festival by eating *tangyuan*, **sticky** rice balls, meaning happiness and reunion, and this festival marks the end of the Spring Festival.

The Spring Festival is a great time for family reunion, so people make great efforts to return home from far places to reunite with their families, which tends to result in **the Spring Festival travel rush**. Besides, the festival mirrors people's desire for a promising future.

The Qingming Festival

The Qingming Festival usually falls on the 4th, 5th or 6th day of April. It is the best time for spring plowing and sowing as the weather turns warmer and the rainfall becomes more plentiful. Unlike other traditional festivals in China, the Qingming Festival is a very special one, which distinguishes itself with its dual identity. It is not only a solar term to remind people of climate change and farming activities, but also a festival in commemoration of the **deceased**. And it is closely related to the **Cold Food Day**, a festival to **commemorate Jie Zitui**[2].

When the Qingming Festival comes, families will pay a solemn visit to the tombs of their ancestors to show their sincere affection and respect by sweeping tombs, offering some **sacrifices** and burning **joss paper**. While with the development of society, the forms of sacrifice have become more environment-friendly. Presenting flowers to the deceased are preferred by a growing number of people nowadays.

The festival is also a good time for people to relax by participating in some activities beneficial to physical and mental health. People like to go for a walk in the countryside with their families or friends, **indulging** themselves in the beautiful spring scenery. People also enjoy themselves with other activities such as playing football, **tug of war** and swings and flying kites.

Kite

The Qingming Festival reflects people's **filial piety** to their forefathers, which has been considered a key virtue in Chinese culture. In addition, the activities of going for **outings** in spring mirror the harmonious relationship between human and nature.

The Dragon Boat Festival

The Dragon Boat Festival is a time-honored traditional Chinese festival on the 5th day of the 5th month. There are definitely various versions about the origin of the Dragon Boat Festival. Yet the most influential and convincing one is to honor **Qu Yuan**[3], a great **patriotic** poet and a high official in the Chu State during the Waring States Period. Qu Yuan was wronged and **banished** by the king of Chu from the court. Terribly frustrated, he committed suicide by throwing himself into **Miluo River**[4]. The local people were very sad after getting the news, because they admired him for his love of the country and people. They rowed boats in an attempt to save him but failed, but the boat-racing has been passed down. Today, people still have dragon boat races to commemorate him on the 5th day of the 5th month every year. It has become the grandest activity on the festival, **prevailing** in the south of China.

Legend has it that people attempted to save his body from fish by throwing eggs and rice wrapped up in bamboo leaves into Miluo River. This graduated into the custom of eating *zongzi* on this festival. Today the custom has spread to Korea, Japan and the countries in Southeast Asia. Other activities include wearing perfume bags （香包）, winding colorful threads around children's wrists and hanging wormwood （苦艾） on the doors. All these have a common purpose to keep the evil spirit away.

The Dragon Boat Festival **embodies** a profound sense of patriotism of Qu

Yuan, which has been advocated until now. Furthermore, the festival also witnesses the harmonious relationship between human and nature, shown by the various activities of dragon boat races, eating *zongzi*, etc.

The Double Seventh Festival

As Chinese Lover's Day, the Double Seventh Festival is celebrated on the 7th day of the 7th month. It is not the most well-known but the most romantic one.

Its legend goes like this: Once upon a time, a beautiful fairy in heaven fell in love with a herd boy on earth. She was so devoted to her love that she gave up all her **privileges** in heaven and came down on earth to live with her beloved. Ever since then they lived in love and harmony. The Herd boy worked as hard as usual in the field while the Weaving Girl raised silkworms and wove **exquisite** silks which won her great reputation. Three years

The Weaving Girl and the Herd

later, they had two children and lived a much happier life than before. But when her mother, the **Queen Mother**[5], learned about this she was so angry that she had her daughter back to heaven by force. The Herd Boy, with their two children, ran after her wife, but the Queen Mother took out her **hairpin** and drew a line between them. The line became a river immediately, separating the couple. To help them, thousands of **magpies** formed a bridge across the river so that they could meet each other once a year. Because the Weaving Girl was a symbol of beauty, kindness and **ingenuity**, girls would pray to her to gift them intelligence and ingenious skills on the eve of this festival.

This story casts a romantic light on this festival. In fact, it demonstrates people's **yearning** for free love. Meanwhile, it also portrays the ancient agricultural civilization in which men plowed fields and women wove cloth.

The Mid-Autumn Festival

The Mid-Autumn Festival falls on the 15th day of the 8th month, featuring a full-moon night and marking exactly the middle of autumn. The full moon is considered as a symbol of reunion. Therefore, it is also a festival for family reunion.

The Mid-Autumn Festival has a long history, which can be traced back to Han Dynasty. During that period, offering sacrifices to the moon and enjoying the moonlight were very important customs. It became an **established** festival during Song Dynasty. As time goes by, its religious and **ritual** implications have gradually disappeared, while the significance of family reunion has been strengthened.

The festival is celebrated mainly in the following two ways. The first is offering sacrifices to the moon and enjoying the moon. The second is eating moon cakes. Like eating *zongzi* on the Dragon Boat Festival and eating *tangyuan* on the Lantern Festival, eating moon cakes is a traditional Chinese custom on this festival. The moon cake is usually round, symbolizing reunion. On the Mid-Autumn Festival, the family will get together, eat moon cakes, enjoy the bright moon and chat leisurely.

Like the Spring Festival, the Mid-Autumn Festival embodies the cultural connotation of family reunion. This explains why the family members enjoy a splendid feast with moon cakes as the major desserts. Meanwhile, the custom of worshiping and enjoying the moon shows the harmony between human and nature.

The above festivals open a window to the traditional Chinese culture. They carry a lot of cultural meanings, such as people's wishes for family reunion, harmony between human and nature, inheritance of filial piety, commemoration of their ancestors and patriotic heroes and so on. They nourish Chinese people's spirits of the **reverence** for nature and love for peace. As a result, the traditional festivals need not only to be preserved but also to be inherited. In addition, it is also necessary to introduce the festivals to foreigners, thus they are able to get a better understanding of Chinese culture.

New words ...

nourishment [ˈnʌrɪʃmənt] *n.* If something provides a person, animal, or plant with nourishment, it provides them with the food that is necessary for life, growth, and good health. 滋养品；营养

connotation [ˌkɒnəˈteɪʃn] *n.* The connotations of a particular word or name are the ideas or qualities which it makes you think of. （词或名字的）内涵意义

evolve [iˈvɒlv] *v.* If something evolves or you evolve it, it gradually develops over a period of time into something different and usually more advanced. （使）逐步发展；（使）演化

demon [diːmən] *n.* A demon is an evil spirit. 魔鬼；恶魔

sticky [ˈstɪki] *adj.* A sticky substance is soft, or thick and liquid, and can stick to other things. Sticky things are covered with a sticky substance. 黏性的；涂有黏性物质的

deceased [dɪˈsiːst] *n.* A deceased person is one who has recently died. 死者

commemorate [kəˈmeməreɪt] *v.* To commemorate an important event or person means to remember them by means of a special action, ceremony, or specially created object. 纪念

sacrifice [ˈsækrɪfaɪs] *n.* Giving sacrifices means you give up something important or valuable in order to get or do something that seems more important. 牺牲；舍弃

indulge [ɪnˈdʌldʒ] *v.* If you indulge in something or if you indulge yourself, you allow yourself to have or do something that you know you will enjoy. 纵情享受；沉溺于

filial [ˈfɪliəl] *adj.* You can use filial to describe the duties, feelings, or relationships which exist between a son or daughter and his or her parents. 孝顺的

piety [ˈpaɪəti] *n.* Piety is strong religious belief, or behaviour that is religious or morally correct. 虔诚；虔敬

outing [ˈaʊtɪŋ] *n.* An outing is a short enjoyable trip, usually with a group of people, away from your home, school, or place of work. （通常指集体）远足；短途旅行

patriotic [ˌpætrɪˈɒtɪk] *adj.* Someone who is patriotic loves their country and feels very loyal towards it. 爱国的

banish [ˈbænɪʃ] *v.* If someone or something is banished from a place or area of activity, they are sent away from it and prevented from entering it. 放逐

prevailing [prɪˈveɪlɪŋ] *adj.* Prevailing means most frequent or common at a particular time. 流行的；普遍的；盛行的

embody [ɪmˈbɒdi] *v.* To embody an idea or quality means to be a symbol or expression of that idea or quality. 表现；体现；代表（思想或品质）

privilege [ˈprɪvəlɪdʒ] *n.* If you talk about privilege, you are talking about the power and advantage that only a small group of people have, usually because of their wealth or their high social class. （常指因财富、社会地位而享有的）特权；优惠

exquisite [ɪkˈskwɪzɪt] *adj.* Something that is exquisite is extremely beautiful or pleasant, especially in a delicate way. 精美的；精致的

hairpin [ˈheəpɪn] *n.* A hairpin is a small piece of metal or plastic bent back on itself which someone uses to hold their hair in position. 小发夹；小发针

magpie [ˈmæɡpaɪ] *n.* A magpie is a large black and white bird with a long tail. 喜鹊

ingenuity [ˌɪndʒəˈnjuːəti] *n.* Ingenuity is skill at working out how to achieve things or skill at inventing new things. 足智多谋；心灵手巧

yearning [ˈjɜːnɪŋ] *n.* A yearning for something is a very strong desire for it. 渴望；强烈愿望；向往

established [ɪˈstæblɪʃt] *adj.* If you use established to describe something such as an organization, you mean that it is officially recognized or generally approved of because it has existed for a long time. 已被认可的；已被接受的

ritual [ˈrɪtʃuəl] *n.* A ritual is a religious service or other ceremony which involves a series of actions performed in a fixed order. 典礼；（宗教）仪式

reverence [ˈrevərəns] *n.* Reverence for someone or something is a feeling of great respect for them. 尊敬；崇敬；敬意

Phrases and expressions

pass away　去世
fall on　落到
engage oneself in　从事于
in commemoration of　纪念
go for　对某事物有兴趣；爱好
commit suicide　自杀
pass down　把…一代传一代；使流传
graduate into　渐渐变为
cast light on　阐明；使人了解
trace back to　追溯到
bestow upon　赠予

Phrases and expressions

the Spring Festival travel rush　春运高峰
Cold Food Day　寒食节
joss paper　民间祭祀时用以礼鬼神和葬礼及扫墓时用以供死者享用的纸钱
tug of war　拔河比赛

Chapter 6 Traditional Festivals and Folk Customs

Cultural Notes ...

1. Jade Emperor is the Taoist ruler of Heaven and all realms of existence below including that of Man and Hell according to a version of Taoist mythology. He is one of the most important gods of the Chinese traditional religion pantheon.

2. Jie Zitui was a Han aristocrat who served the Jin prince Chong'er during the Spring and Autumn Period of Chinese history. Chinese legend holds that when Chong'er finally ascended to power as the duke of Jin, Jie either refused or was passed over for any reward, despite his great loyalty during the prince's times of hardship. Supposedly, the duke so desired to repay Jie's loyalty that, when Jie declined to present himself at court, he ordered a forest fire to compel the recluse out of hiding. Instead, Jie and his mother were killed by the fire on Mt Mian. He was annually commemorated with a ritual avoidance of fire that, despite many official bans, eventually became China's Cold Food and Qingming Festivals.

3. Qu Yuan was a Chinese poet and minister who lived during the Warring States period of ancient China. He is known for his patriotism and contributions to classical poetry and verses, especially through the poems of the Chu Ci anthology. During the early days of King Huai's reign, Qu Yuan was serving the State of Chu as its Left Minister. However, King Huai exiled Qu Yuan to the region north of the Han River, because corrupt ministers slandered him and influenced the king. Eventually, Qu Yuan committed suicide by wading into the Miluo River in today's Hunan Province while holding a rock.

4. Miluo river is located on the eastern bank of Dongting Lake, the largest tributary of the Xiang River in the northern Hunan Province. It is an important river in the Dongting Lake watershed, known as the location of the ritual suicide in 278 BC of Qu Yuan, a poet of Chu state during the Warring States period, in protest against the corruption of the era.

5. Queen Mother, known by various local names, is a goddess in Chinese religion and mythology, also worshiped in neighbouring Asian countries, and attested from ancient times. She is most often associated with Taoism. Zhuangzi describes the Queen Mother as one of the highest of the deities, meaning she has gained immortality and celestial powers.

Exercises

1. Read the following statements and try to decide whether it is true or false according to your understanding.

____ 1) Visiting relatives and friends is an indispensable part of the Spring Festival.

____ 2) The Dragon Boat Festival is a traditional festival for people to commemorate their deceased ancestors.

____ 3) According to historical records, Qu Yuan was born in a quite poor family in the Chu State.

____ 4) Valentine's Day is a traditional Chinese festival to show love to others.

____ 5) The Mid-Autumn Festival has had a long history, which can be traced back to Han Dynasty.

2. Fill in the blanks with the information you learn from the text.

1) Among traditional Chinese festivals, _____ is regarded as the most important festival by the Chinese people.

2) Originated from the legend of Qu Yuan, eating _____ is an important tradition among Chinese people.

3) _____ is Chinese Lover's Day, a romantic festival among Chinese festivals.

4) As a family reunion occasion, the Mid-Autumn Festival is also called _____.

5) Eating _____ on the Mid-Autumn Festival is a traditional Chinese custom.

3. Explain the following terms.

   1）The Dragon Boat Race
   2）The Moon Cake

4. Translate the following paragraph into English.

春节是中国民间最隆重、最富有特色的传统节日。一般指除夕和正月初一，是一年的第一天，又叫阴历年，俗称"过年"。在春节期间，我国的汉族和很多少数民族都要举行各种活动以示庆祝。这些活动均以祭祀神佛、祭奠祖先、除旧布新、迎禧接福、祈求丰年为主要内容。活动内容丰富多彩，带有浓郁的民族特色。

5. Critical thinking and discussion.

Western festivals have been increasingly popular in our country in recent years. Some people hold that those western festivals have promoted the exchanges of Chinese and western cultures. While others believe that they have been culturally distressing to Chinese young people. What is your opinion on it?

Section B Marital and Funeral Customs

Marital Customs

Ever since ancient times, one popular saying in China has been going that there are four happiest events in one's life: to have a good rain after a long **drought**, to run across an old friend in a distant land, to pass the examination and to spend wedding night. Marriage is listed one of them for its significance in one's life.

With time going by, a lot of **marital** customs have been established. Despite the great changes in different places, some of the basic customs have been passed down and still observed today. A wedding is usually a grand occasion with overly elaborate ceremonies. In China, the most essential and popular rituals are known as *liuli*（六礼）: proposing marriage, matching birth dates, **submitting** engagement gifts, presenting wedding gifts, selecting wedding date and holding wedding ceremony.

Proposing Marriage. In the past, a marriage was arranged by parents. Once a girl was selected as an ideal match by the unmarried boy's parents or his matchmaker, the boy's parents would invite the matchmaker to formally present a request to the girl's parents.

Matching Birth Days. If the girl's parents approved of the marriage, the matchmaker would then ask for the girl's birthday and birth hour to assure the **compatibility** of the girl and the boy.

Submitting engagement Gifts. If their birth hours did not conflict according to **astrology**, the boy's family would ask the matchmaker to send some initial gifts to the girl's family.

Presenting Wedding Gifts. The boy's family would pick a **white day** and send the girl's family the bridal gifts such as tea, wine and jewelry, depending on local customs and household wealth. This act **confirmed** the marriage agreement between the two families, which is quite similar to the current **engagement.**

Selecting Wedding Day. The boy's parents would consult a **fortune teller** to select a white day according to the birth dates of the boy and the girl and their family. The girl's family would prepare **dowry** for her. The dowry, indicating the girl's family social status and wealth, would be on display.

Holding Wedding Ceremony. There is no doubt that the wedding ceremony is the most significant and interesting one among *liuli*. Before the wedding day, both the bride and the groom's houses would be decorated in red and pasted with happy Chinese cha-racters: **double happiness**[1]. On the wedding day, the groom's family would send out a procession of friends, musicians and a carriage to the bride's family to greet the bride. With firecrackers exploding and **band** playing, the bride in red would then be taken to the groom's house and the two would perform the marriage ceremony witnessed by all the relatives and friends. The bride and groom would perform formal bows. The ritual was also called **Bowing**

Double Happiness

to Heaven and Earth[2]（拜天地）, which was conducted to gain approval of the marriage from gods, ancestors and parents as well.

After that, the groom's family would then hold a huge feast for the friends and relatives to celebrate the wedding. After the meal, the newly-married couple would return to the bridal **chamber**. Their wedding bed had been **sprinkled** with red dates, peanuts, dried longan and lotus seeds, which in Chinese means giving birth to a baby as early as possible. The couple would drink **cross-cupped wine**, symbolizing their **integration**. And the wedding guests may play tricks on the bride and groom and they wouldn't leave till the dead of night. Then, the groom would take off **the red veil** covering the bride's face. Thus, the wedding ceremony finished.

Traditional Wedding Ceremony

Traditional Chinese marital customs have changed a great deal today. A lot of old complicated customs are abandoned, and new forms of wedding become popular. Young people are able to arrange their marriage and wedding considerably freely. An increasing number of couples choose a wedding trip and even a mass wedding. However, in recent years, the traditional wedding ceremony has been **reviving** because of its attraction to some young people.

The marital customs reflect that Chinese people attach great importance to family. As a basic **societal** building block, family has traditionally been the cornerstone of society. Consequently, **carrying on the family line** has been regarded as an important thing. It is only after marriage that the bride and groom can legally establish a family and bear children to continue the family line. Wedding ceremony adds a ritual sense to this big event in one's life.

Modern Wedding Ceremony

Funeral Customs

Funeral customs have always been viewed as an important part of Chinese social life on account of great importance attached to death in our country. A series of rituals are performed from one's death to funeral. Chinese funeral rituals comprise a set of traditions associated with Chinese folk religions, with different **rites** depending on the age, cause of death and social status of the deceased. Different rituals are carried out in different places of China. However, despite many variations, there is a set of standardized and **prescribed** ritual behaviors. The main rituals are as follows:

Attending upon a Dying Parent (送终). When a parent is dying, all the family members are summoned to gather around the bedside to say farewell to him. It is considered **unfilial** not to be at the deathbed of one's parent. At the moment of death, the family members will burst into loud **wailing** and crying.

Washing and Clothing the Deceased (小殓). The deceased needs to be cleaned with a towel and dressed in his best clothes. Such clothes are usually white, black, blue or brown. All other clothes that the deceased used to wear when he was alive are all burnt or thrown away.

Coffining (大殓). After being cleaned and clothed, the deceased is placed into a coffin. Some personal items of the deceased are also placed inside it, in the belief that he will continue to use them in the **afterworld**. **Wreaths**, food, fruit and a photograph of the deceased are placed in front of the coffin. At the foot of the coffin is an **altar**, upon which burning **incense** and a lighted white candle are put.

Wreath

Holding the Funeral Wake (守灵). During the wake, the family members do not wear jewelry or red clothing for the reason that red is the color of happiness. Instead, they wear special white garments as a sign of mourning. The family members gather around the coffin. Traditionally, the family of the deceased needs to keep all-night **vigils**, which is a way to show filial **piety** and loyalty to the deceased. The funeral wakes range from three to seven days, enabling relatives and friends to pay their last respect to the deceased. On arriving at the home of the deceased, the visitors in **somber** colors will perform three ritual bows to the deceased. One or more members of the deceased's family will stand or kneel by the side of the altar to bow to the visitors for respect.

Being Buried（出殡）. On the day of being buried, the coffin is put on the **hearse** which will take it to the place of burial. The funeral procession follows the hearse, with sons and daughters in the first row, followed by other family members. Paper figures of cars, ships or other objects are sometimes carried as necessities of the deceased in the afterworld. On arrival at the graveyard, the coffin is lowered into the ground. Family members and friends pay their final respect to the deceased. After the burial, family members will wear a piece of black cloth on one of their sleeves for a short period to show their **mourning**. In the past, traditional **inhumation** was favored, while at present the deceased are often **cremated**, particularly in large cities in China.

The funeral customs are the perfect reflection of filial piety, the virtue of respect to one's parents, elders and ancestors in Confucian philosophy. Filial piety has been considered a great virtue in Chinese culture and has been held in high **esteem** for thousands of years. Consequently, a series of funeral rituals are carried out by family members to show their filial piety to the deceased. However, such funeral rituals cause a huge waste in terms of time, energy and expenses of families of the deceased. Therefore, it is necessary to consider how to simplify traditional funeral customs in a respectful way.

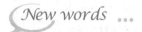

drought [draʊt] *n.* A drought is a long period of time during which no rain falls. 久旱；旱灾

marital [ˈmærɪtl] *adj.* Marital is used to describe things relating to marriage. 婚姻的

submit [səbˈmɪt] *v.* If you submit a proposal, report, or request to someone, you formally send it to them so that they can consider it or decide about it. 提交；递呈（建议、报告或请求）

compatibility [kəmˌpætəˈbɪləti] *n.* Compatibility refers to the ability of people or things to live to exit together without problems. 并存；相容

astrology [əsˈtrɒlədʒi] *n.* Astrology is the study of the movements of the planets, sun, moon, and stars in the belief that these movements can have an influence on people's lives. 占星术

confirm [kənˈfɜːm] *v.* If something confirms what you believe, suspect, or fear, it shows that it is definitely true. 证实；证明（情况属实）

engagement [ɪnˈɡeɪdʒmənt] *n.* An engagement is an agreement that two people have made with each other to get married. 婚约

dowry [ˈdaʊri] *n.* A woman's dowry is the money and goods which, in some cultures, her family gives to the man that she marries. 嫁妆；陪嫁

band [bænd] *n.* A band is a small group of musicians who play popular music such as jazz, rock, or pop. 乐队；乐团

chamber [ˈtʃeɪmbə(r)] *n.* A chamber is a room designed and equipped for a particular purpose. （作特殊用途的）房间

sprinkle [ˈsprɪŋkl] *v.* If something is sprinkled with particular things, it has a few of them throughout it and they are far apart from each other. 使…上零星分布（着…）

integration [ˌɪntɪˈɡreɪʃn] *n.* If someone integrates into a social group, or is integrated into it, they behave in such a way that they become part of the group or are accepted into it. （使）加入；（使）融入

reviving [rɪˈvaɪvɪŋ] *adj.* When something such as the economy, a business, a trend, or a feeling is revived or when it revives, it becomes active, popular, or successful again. （使）复苏；（使）复兴；（使）恢复；（使）再次流行

societal [səˈsaɪətl] *adj.* Societal means relating to society or to the way society is organized. 社会的

rite [raɪt] *n.* A rite is a traditional ceremony that is carried out by a particular group or within a particular society. （传统的）仪式；典礼

prescribe [prɪˈskraɪb] *v.* If a person or set of laws or rules prescribes an action or duty, they state that it must be carried out. 规定；法定

unfilial [ˌʌnˈfɪljəl] *adj.* An unfilial person is the one who ignores the duties, feelings, or relationships which exist between a son or daughter and his or her parents. 不孝的

wail [weɪl] *v.* If someone wails, they make long, loud, high-pitched cries which express sorrow or pain. 哀号；悲鸣；恸哭

afterworld [ˈɑːftəwɜːld] *n.* Afterworld refers to the place where people are after they die. 阴间；阴世

wreath [riːθ] *n.* A wreath is an arrangement of flowers and leaves, usually in the shape of a circle, which you put on a grave or by a statue to show that you remember a person who has died or people who have died. 花圈

altar [ˈæltə] *n.* An altar is a holy table in a church or temple. 祭坛；神坛

incense [ˈɪnsens] *n.* Incense is a substance that is burned for its sweet smell, often as part of a religious ceremony. （常用于宗教仪式的）香

vigil [ˈvɪdʒəl] *n.* A vigil is a period of time when people remain quietly in a place, especially at night, for example because they are praying or are making a political protest. 守夜

piety [ˈpaɪɪti] *n.* Piety is strong religious belief, or behavior that is religious or morally correct. 虔诚；虔敬

somber [ˈsɒmbə(r)] *adj.* If someone is somber, they are serious or sad. 忧郁的；严肃的

hearse [hɜːs] *n.* A hearse is a large car that carries the coffin at a funeral. 灵车；柩车

mourning [ˈmɔːnɪŋ] *n.* Mourning is behavior in which you show sadness about a person's death. 哀悼；伤逝

inhumation [ˌɪnhjuːˈmeɪʃn] *n.* Inhumation is the ritual placing of a corpse in a grave. 土葬

cremate [krə'meɪt] *v.* When someone is cremated, their dead body is burned, usually as part of a funeral service. 焚烧；火化（尸体）

esteem [ɪ'sti:m] *n.* Esteem is the admiration and respect that you feel towards another person. 尊敬；尊重；敬重

Phrases and expressions

on account of　由于
approve of　同意
on display　展示；展出
play tricks on　捉弄；戏耍

Proper names

a white day　良辰吉日，常用于称宜于成亲的日子。
fortune teller　算命先生，一种给人算命的职业称谓。
cross-cupped wine　交杯酒，传统婚俗之一，源于先秦。新郎新娘进入洞房后先各饮半杯，然后交换一齐饮干，谓之饮交杯酒，在古代又称为"合卺"。
the red veil　古时候婚礼时，新娘头上都会蒙着一块别致的大红绸缎，被称为红盖头，这块盖头要入洞房时由新郎揭开，是民间迎亲途中的礼仪之一。
carrying on the family line　传宗接代

Chapter 6 Traditional Festivals and Folk Customs

Cultural Notes ...

1. Double happiness, sometimes translated as Double Happy, is a Chinese traditional ornament design, commonly used as a decoration and symbol of marriage. Double Happiness is a blessing traditionally bestowed upon a Chinese bride and groom, meaning "May you have twice as much happiness and half as much sadness in your marriage".

2. Bowing to Heaven and Earth is a very important ceremony in wedding, indicating a good start in the marriage. During the ceremony, the person presiding over the wedding will say loudly: "First, bow to heaven and earth; second, bow to your parents; third, bow to husband and wife, and then enter the bridal chamber, please." Today, instead of bowing to Heaven and Earth, bowing to parents is more often performed.

Exercises ...

1. Read the following statements and try to decide whether it is true or false according to your understanding.

____ 1) The wedding customs have gone through great changes with time going by.

____ 2) In the past, young girls and young boys were able to find their future spouse at their own will.

____ 3) Before the wedding ceremony, the bride's family would need to prepare dowry for the girl.

____ 4) It is considered unfilial not to be at the deathbed of one's parent.

____ 5) During the funeral wake, the family members can wear jewellery or red clothing.

2. Fill in the blanks with the information you learn from the text.

1) Ever since ancient times, a popular saying has been going in China that there are four happiest events in one's life: to have a good rain after a

long drought, to run across an old friend in a distant land, to pass the examination and to spend _____.

2) Once an unmarried boy's parents planned to select a girl as their ideal future daughter-in-law, they would request a _____ who would find a suitable girl for the boy.

3) There is no doubt that _____ is the most significant and interesting one among *liuli*.

4) After being cleaned and clothed, the deceased is placed into a _____.

5) After the burial, family members will wear _____ on one of their sleeves for a short period to show their mourning.

3. Explain the following terms.

1) *Liuli*
2) Holding the funeral wake

4. Translate the following paragraph into English.

"拜堂",又称为"拜天地",是婚礼中一个很重要的仪式。"拜堂"这一婚俗于宋代以后非常流行,经过"拜堂"后,女方就正式成为男家的一员。"拜堂"时,主持婚礼的司仪会大声地说:"一拜天地,二拜高堂,夫妻交拜,齐入洞房。"其实,拜天地代表着对天地神明的敬奉;而拜高堂就是对孝道的体现;至于夫妻交拜就代表夫妻相敬如宾。

5. Critical thinking and discussion.

Funeral rituals have always been viewed as an important part of Chinese social life for the reason that it reflects Chinese culture of filial piety. Usually a funeral takes at least three or four days, involving a variety of complex rituals, consuming much time, energy and money of the family of the diseased. Do you think it is advisable to simplify the funeral rituals? And give your own reasons for your opinion.

Section C Temple Fair

Temple fair is a folk activity that is developed from religious activities held

Chapter 6   Traditional Festivals and Folk Customs

in temples. It is a part of Chinese traditional culture and a kind of precious tourism resource. Its emergence, existence and development are closely related to the life of ordinary people.

Origin and Development of Temple Fair

Temple Fair

Legend has it that temple fair originated in ancient times when people offered sacrifices to gods in a temple. The early temple fair was only a grand sacrifice. During the period of holding sacrificial activities, some traders and **peddlers** set up various **stands** outside the temple to earn money from people who came to burn **incense** and worship gods inside. Thus the temple location gradually evolved into a marketplace for people to exchange products as well as a place for cultural performances. With the development of temple fair for a long time, it has gradually become a local folk custom and a mass gathering integrating religious worship, commercial activity and entertainment. Now temple fair is an important and joyful destination for Chinese people during festivals, especially the Spring Festival. Some temple fairs are held regularly, and others are held on the occasions when there are traditional festivals or something related to sacrificial activities for Buddhism, Taoism and other religions.

Activities at Temple Fair

There are mainly three kinds of activities at temple fair: worship ceremonies, artistic folk performances and commercial activities.

Actually, the customs of temple fair are closely related to the religious activities of Buddhist and Taoist temples. Therefore, there has been a long tradition to hold worship ceremonies at temple fair. There are many kinds of worship ceremonies, among which the "tour of inspection" (巡城) is a typical one. That is, the statues of the gods in the temple are carried out and followed by both the **deacons** and faithful men and women dressed up in various costumes. Worship ceremony becomes less popular today. Despite that, a large number of people still go to the temple to burn incense and pray to gods with a desire to seek good fortune and avoid disaster.

Stilts Performance

Various artistic folk performances are on display at temple fair. The following activities are performed for the enjoyment of the visitors: land boat (划旱船), lion dance, drum dance, gong (锣) and drum beating, **stilts** walking, **acrobatics**, magic shows, operas, **cross-talk shows**, *yangge*[1] dance and so on. In fact, some traditional performances such as cross talks, *kungfu* shows and lion dances have become the major features of the fair. Besides, some special craft skills are also displayed, such as sugar painting, sugar-figure blowing (吹糖人), shadow puppetry (皮影戏), egg carving, **embroidery**, paper cutting, etc.

Temple fair is actually a giant market, attracting a large number of merchants, peddlers and consumers. On the days of temple fair, thousands of goods are on sale, including jewelry, silk, clothing, food and drinks, paintings, **calligraphic** works, **antiques**, articles for daily use, children's toys, and seasonal fruit and vegetables, etc. It's no exaggeration that all kinds of goods are available. Among the goods, various tasty **snacks** are the most attractive ones. Tourists can treat themselves to delicious food from almost all over the country. In addition, a good range of **fancy** toys are also sold, dazzling both adults and children.

There are a large number of temple fairs in China, among which Beijing Ditan Temple Fair, Shenyang Royal Temple Fair, Shanghai City Temple Fair and **Nanjing Confucius Temple**[2] Fair are the four **top-notch** ones. Take Beijing Ditan Temple Fair for example. Ditan Park is the site of the altar where sacrifices were formerly offered to the **God of Earth**[3]. Based on sacrifices in the park, Ditan Temple Fair gradually emerged and developed. At present, it has been one of the most famous temple fairs. Similar to other temple fairs, during the Spring Festival, Ditan Temple Fair is characterized by worship ceremonies, artistic folk performances and commercial activities. Traditional Qing ceremonies are **reenacted** to worship the God of Earth for blessing every year. At the temple fair there is a unique square for **Intangible Cultural Heritage,** offering the performances of **pottery** making, paper cutting, shadow puppetry, homespun weaving, etc. Besides, there are also various interactive competitions such as **arm-wrestling,** rock climbing and chess playing. As an **indispensable** part of the fair, numerous goods are available, including paintings, calligraphic works and other goods. There is no denying that snacks are quite **appealing** to visitors. They can taste a variety of local snacks in Beijing, especially Beijing roast ducks and court food. Ditan Temple Fair is also popular among foreigners who are able to have access to Chinese traditional culture by appreciating **craftsmanship** and artwork displayed by local artisans at the fair.

The Worship Ceremony at Ditan Temple Fair

  Temple fair is an important component of traditional **folklore** and social life in China. The **bustling** temple fair is an ideal place for people to entertain themselves. It offers a precious opportunity for people to enjoy different kinds of artistic folk performances, among which many traditional folk arts cannot be enjoyed in daily life, and even some of them are in danger of being lost. So it is no exaggeration that temple fair plays a vital role in inheriting folk art in China. Therefore, temple fair should be well planned and further normalized to function as a better stage to show traditional Chinese culture.

New words

peddler ['pedlə(r)] *n*. A peddler is a person who goes from place to place in order to sell something. 流动小贩；货郎

stand [stænd] *n*. A stand is a small shop or stall, outdoors or in a large public building. （设在户外或大型公共建筑物内的）小店；售货摊

incense ['ɪnsens] *n*. Incense is a substance that is burned for its sweet smell, often as part of a religious ceremony. 香

deacon ['diːkən] *n*. A deacon is a member of the clergy, for example in the Church of England, who is lower in rank than a priest. 执事

stilt [stɪlt] *n*. Stilts are two long pieces of wood with pieces for the feet fixed high up on the sides so that people can stand on them and walk high above the ground. 高跷

acrobatics [ˌækrəˈbætɪks] *n*. Acrobatics are acrobatic movements. 杂技

embroidery [ɪmˈbrɔɪdəri] *n*. Embroidery consists of designs stitched into cloth. 绣花；刺绣品

calligraphic [ˌkælɪˈɡræfɪk] *adj*. Being calligraphic is related to or expressed in calligraphy. 书法的

antique [ænˈtiːk] *n*. An antique is an old object such as a piece of china or furniture which is valuable because of its beauty or rarity. 古物；古玩；古董

snack [snæk] *n*. A snack is something such as a chocolate bar that you eat between meals. 小吃；零嘴

fancy [ˈfænsi] *n*. If you describe something as fancy, you mean that it is special, unusual, or elaborate, for example because it has a lot of decoration. 别致的；精美的

top-notch [tɒp nɒtʃ] *adj*. If you describe someone or something as top-notch, you mean that they are of a very high standard or quality. 一流的；顶尖的

reenact [riːˈnækt] *v*. If you re-enact a scene or incident, you repeat the actions that occurred in the scene or incident. 再次展现；再现；重现

pottery [ˈpɒtəri] *n.* You can use pottery to refer to pots, dishes, and other objects which are made from clay and then baked in an oven until they are hard. 陶器；陶瓷器

arm-wrestling [ˈɑːmˌreslɪŋ] *n.* 掰手腕

indispensable [ˌɪndɪˈspensəbl] *adj.* If you say that someone or something is indispensable, you mean that they are absolutely essential and other people or things cannot function without them. 必不可少的；不可或缺的

craftsmanship [ˈkrɑːftsmənʃɪp] *n.* Craftsmanship is the skill that someone uses when they make beautiful things with their hands. 手艺；技艺

folklore [ˈfəʊklɔː(r)] *n.* Folklore is the traditional stories, customs, and habits of a particular community or nation. 民间传说；民俗

bustling [ˈbʌslɪŋ] *adj.* A place that is bustling with people or activity is full of people who are very busy or lively. 繁忙；活跃

Phrases and expressions

evolve into 逐渐发展成
a good range of 种类繁多
have access to 可以；能够

Proper names

cross-talk shows 相声，一种民间说唱曲艺。它以说、学、逗、唱为形式，突出其特点。
Intangible Cultural Heritage 非物质文化遗产

Cultural Notes

1. *Yangge* is one of the traditional folk dances in China. It is usually performed in northern provinces. *Yangge* has a long history, originated in the working life of planting and farming. Farmers worshiped Gods to pray for bumper harvests, and away from disasters with singing and dancing. Gradually the ceremonies developed into *yangge* dance.

2. Nanjing Confucius Temple is a place to worship and consecrate Confucius, a great philosopher and educator of ancient China. It is also known as Fuzimiao in Chinese. Originally, the temple was constructed in the year of 1034 in Song Dynasty. It suffered repeated damage and has been rebuilt on several occasions since that time. In front of the Confucius Temple, the Qin Huai River is flowing. Around the temple are a series of tourist shops, snack bars, restaurants and tea cafes. They all appear to be in the architectural style of the Ming and Qing style.

3. God of Earth, also known as Lord of the Soil and the Ground, is a deity and a local protector, worshiped in Chinese folk religion. In Chinese traditional culture, offering sacrifices to the God of Earth means sacrificing the earth. However, in modern times the sacrifices mainly mean praying for blessings, safeguarding peace and protecting the harvest.

Exercises

1. Read the following statements and try to decide whether it is true or false according to your understanding.

____ 1) The emergence of the temple fair was closely related to religion.

____ 2) The original temple fairs already integrated religious worship, entertainment and commercial activity.

____ 3) There are only a few kinds of artistic folk performances at the temple fair.

____ 4) The goods sold at the temple fair are so rich in types that we can say almost all the goods are available.

Chapter 6   Traditional Festivals and Folk Customs

_____ 5) Ditan Temple Fair is not popular among foreigners living and working in Beijing.

2. Fill in the blanks with the information you learn from the text.

1) With the development of temple fairs, gradually they have become a mass gathering integrating religious worship, entertainment and _____.

2) On the temple fair, there are various artistic folk performances for _____.

3) Among the goods sold on the temple fair, _____, with a wide kind, are the most attractive ones.

4) _____, together with Shenyang Royal Temple Fair, Shanghai City Temple Fair and Nanjing Confucius Temple Fair constitute the four top-notch temple fairs in China.

5) Ditan Temple Fair features reenactments of traditional Ming and Qing ceremonies to worship the _____.

3. Explain the following terms.

1) Artistic Folk Performances at Temple Fair
2) Ditan Temple Fair

4. Translate the following paragraph into English.

庙会是中国民间宗教风俗，一般在农历新年、元宵节举行。庙会也是中国集市贸易形式之一，其形成与发展和地庙的宗教活动有关，在寺庙的节日或规定的日期举行，多设在庙内及其附近，进行祭神、娱乐和购物等活动。庙会流行于全国广大地区，也是中国民间广为流传的一种传统民俗活动。

5. Critical thinking and discussion.

Temple fair is an organic part of the Chinese folk custom. During the Spring Festival, temple fair has become a good destination for people to relax and enjoy themselves. It has brought about joy, excitement and relaxation to people, meanwhile, it has led to some problems as well, such as the reviving of feudal superstition, disturbing social order and overbuilding temples, etc. In your opinion, what are the effective solutions to regulate the temple fair so as to promote its development?

Chapter 7
Silk and Ceramics Culture

Section A Culture of Silk

"Men **tilling** the land while women weaving at home", a major management mode of the small-scale peasant economy in ancient China, indicates that women, a good match to men in supporting a family, are responsible for weaving to keep the family free of coldness over thousands of years. During the long period of time they have traditionally woven with hemp, silk and cotton. Among them, silk has been playing a vital role in people's social life and well accepted at home and abroad. Even today China's silk products are still popular.

Origin of silk

Leizu

The origin of silk can find its source from different legends, of which two versions are well accepted. One is related to Leizu（嫘祖）, the wife of the Yellow Emperor. One day in the 27th century BC, Leizu was having tea under a mulberry tree when a silkworm cocoon fell into her cup. She examined it and saw a strand of fiber unspun from the cocoon. She tried and found it strong and **flexible** enough to be used for weaving. Then she observed the life of silkworm and grasped the skills of silkworm **breeding** and silk reeling. Then she taught them to her people. And thus began the sericulture（蚕业）in China. Leizu was accordingly honored as the first sericulturist, god of silkworm.

The other is about a love story between a girl and her white horse. Once upon a time, a girl in today's Sichuan Province badly missed her father serving in the army from afar. She promised her white horse a marriage if it could take her father home. However, she ate her words after the horse succeeded in bringing her father back. Heartbroken by her **betrayal**, the horse **whickered** and stopped feeding. His unusual behavior drew the father's attention, so he questioned his daughter and got the whole story. Frightened by the horse's strong determination to marry his daughter, he killed and **skinned** it, and hung the horsehide in the yard. One day, the horsehide wrapped up the girl and fled away. Shortly afterwards they were found to have transformed into a silkworm with horse head and human body in the mulberry wood. That is why the white worm was named Matouniang（马头娘）.

Development of Sericulture

Fragments of primitive loom in **Hemudu Culture**[1] in Yuyao, Zhejiang Province suggested there appeared sericulture activities as early as 4000 BC in China. Silk fabric wrapping the body of a child found in Henan Province proved that the use of silk could date back to 3630 BC.

Shang and Zhou Dynasties. In Shang Dynasty, the filature（缫丝）and weaving skills reached a relatively high level. **Complicated** loom and **figured** silk emerged. Due to the limited quantity of silk products, they were available **exclusively** to the royals and nobles. The Warring States Period **witnessed** the prosperity of silk. The production of silk increased and its products could be accessible to ordinary people. The pattern, weaving, **embroidery** and dying skills were all improved. The silk products **unearthed** from **Mawangdui Han Tomb** are the proof of the advanced craftsmanship and free artistry at that time. In Shang and Zhou Dynasties, the Silk Road was opened, which enhanced the trade of silk products. The opening of the route paved the way for the thriving of silk industry of the successive dynasties.

Han Dynasty. The trade **boomed** along the Silk Road. Silk goods from today's Xi'an City went westward to India, **Persia**, **Arabia** and finally to **Mediterranean** countries **overland**; southeastward to today's Guangzhou City and then were shipped southwestward to **Java**, the west bank of Africa across **Indian Ocean**, and finally to Mediterranean countries across **Red Sea**. After such a long journey, the price of silk **soared** as high as that of gold.

Brocade Unearthed from Mawangdui Han Tomb

Ming and Qing Dynasties. In Ming Dynasty, there appeared some records related to silk industry. The medical classic **Compendium of Materia Medica** scientifically identified and classified the mulberry; the agricultural book **Nongzhen Quanshu expounded** the sericultural production; and the comprehensive work **Tiangong Kaiwu** accounted the silkworm **hybridizing**, weaving and dying techniques at that time. Qing government encouraged people to develop silk industry. Emperor Kangxi set up **customs** at some important cities such as **Jianghai**, **Zhehai**, **Minhai** and **Yuehai** to export silk. And Magistrate Lin Qi（林启）opened **Hangzhou Silkworm Academy** in 1897, training a considerable number of professional craftsmen in silk industry. In addition to government involvement, some individual capital was invested into this sector. Chen Qiyuan（陈启元）established China's first silk filature（缫丝厂）in Guangdong Province in 1872. At that time, China's exports of raw silk and silk goods to foreign countries reached a considerable scale. However, the silk industry in the late Qing Dynasty fell into a dilemma under **exorbitant** taxes and levies, and heavy **dumping** of foreign silk.

As Chinese craftsmen constantly improved the weaving skills, they also created various embroidery techniques. In Qing Dynasty, the four famous embroideries (Su, Xiang, Shu and Yue) enjoyed a high reputation. Among them, Su Embroidery was well known for the **exquisite** designs, elegant colors and unique patterns. The product surface was flat, the rim

Cat (Su Embroidery)

neat, the stitch **fine**, the lines **dense**, and the colors harmonious. Its products fell into three major categories: costumes, decorations for halls and crafts for daily use. Double-sided embroidery is an excellent representative. The typical work is an embroidered cat with bright eyes and **fluffy** hair, looking vivid and lifelike.

Uses of Silk

Silk easily **fades** and **wrinkles**. It is **absorbent** and **inelastic**. However, it still gets the favor of people because it's soft, lightweight and **breathable**.

Silk has practical and decorative values. In ancient times, it used to serve as the currency in business as well as the materials for writing and painting. Up to now, it has still been used for some items of menswear, including shirts, suits, ties as well as pocket squares on suit jackets, and many types of women's clothing, such as *qipao*（旗袍）, wedding gowns, evening gowns, skirts and scarves. Besides, silk's attractive **luster** and **drapability** make it suitable for many furnishing applications. It is often used for wall coverings, screens, **ornaments**, silk fans, bedding and wall hangings.

China is home to silk. It is integrated with etiquette, art and folk customs. In the past the privileged wore it to **signify** their social status, literati used it to compose beautiful lines and spectacular paintings, while ordinary people held sacrificial ceremonies to worship God Silkworm. Nowadays silk has become a tie to strengthen the **diplomatic** relations with foreign countries, and a gift to enhance interpersonal relationship. Owing partly to the superb silk weaving technology and rich cultural **connotation**, China's silk has spread all over the world through the Silk Road. It has **stimulated** Chinese national and cultural confidence, and aroused foreigners' yearning for China.

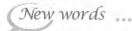

till [tɪl] *v.* When people till land, they prepare the earth and work on it in order to grow crops. 耕；犁（地）

flexible ['fleksəbl] *adj.* A flexible object or material can be bent easily without breaking. 可弯曲的；柔韧的

breed [briːd] *v.* If you breed animals or plants, you keep them for the purpose of producing more animals or plants with particular qualities, in a controlled way. 饲养；培育

betrayal [bɪˈtreɪəl] *n.* A betrayal is an action which betrays someone or something, or the fact of being betrayed. 辜负；背叛；出卖

whicker [ˈwɪkər] *v.* to neigh or whinny（马）嘶

skin [skɪn] *v.* If you skin a dead animal, you remove its skin. 剥去（死动物）的皮；将⋯去皮

complicated [ˈkɒmplɪkeɪtɪd] *adj.* If you say that something is complicated, you mean it has so many parts or aspects that it is difficult to understand or deal with. 复杂的；难懂的；难处理的

figured [ˈfɪɡəd] *adj.* [only before noun] decorated with a small pattern 饰以图案的

exclusively [ɪkˈskluːsɪvlɪ] *adv.* Exclusively is used to refer to situations or activities that involve only the thing or things mentioned, and nothing else. 排他地；独占地；专有地；完全地

witness [ˈwɪtnəs] *v.* If you say that a place, period of time, or person witnessed a particular event or change, you mean that it happened in that place, during that period of time, or while that person was alive. 见证；经历（事件、变化等）

embroidery [ɪmˈbrɔɪdərɪ] *n.* Embroidery consists of designs stitched into cloth. 绣花；刺绣品

unearth [ʌnˈɜːθ] *v.* If someone unearths something that is buried, they find it by digging in the ground. 发掘；挖掘（埋藏物）

boom [buːm] *v.* If the economy or a business is booming, the amount of things being bought or sold is increasing.（经济或生意）景气；繁荣

overland [ˈəʊvəlænd] *adj.* An overland journey is made across land rather than by ship or aeroplane. 经由陆路的；横跨陆地的

soar [sɔː(r)] *v.* If the amount, value, level, or volume of something soars, it quickly increases by a great deal.（数量、价值、水平、规模等）急升；猛涨

expound [ɪkˈspaʊnd] *v.* If you expound an idea or opinion, you give a clear and detailed explanation of it. 详述；阐述；详细说明

hybridize [ˈhaɪbrɪdaɪz] *v.* If one species of plant or animal hybridizes with another, the species reproduce together to make a hybrid. You can also say that you hybridize one species of plant or animal with another. 使（动植物）杂交

customs [ˈkʌstəmz] *n.* Customs is the official organization responsible for collecting taxes on goods coming into a country and preventing illegal goods from being brought in. 海关（部门）

exorbitant [ɪɡˈzɔːbɪtənt] *adj.* If you describe something such as a price or fee as exorbitant, you are emphasizing that it is much greater than it should be. 过度的；过分的；过高的

dump [dʌmp] *v.* If a firm or company dumps goods, it sells large quantities of them at prices far below their real value, usually in another country, in order to gain a bigger market share or to keep prices high in the home market. （通常指为占领海外市场）倾销；抛售

exquisite [ɪkˈskwɪzɪt] *adj.* Something that is exquisite is extremely beautiful or pleasant, especially in a delicate way. 精美的；精致的

fine [faɪn] *adj.* Something that is fine is very delicate, narrow, or small. 纤细的；尖细的；微小的

dense [dens] *adj.* Something that is dense contains a lot of things or people in a small area. 稠密的；密集的

fluffy [ˈflʌfi] *adj.* If you describe something such as a towel or a toy animal as fluffy, you mean that it is very soft. 绒毛似的；松软的；蓬松的

fade [feɪd] *v.* When a colored object fades or when the light fades it, it gradually becomes paler. （使）褪色；（使）变暗淡

wrinkle [ˈrɪŋkl] *v.* If cloth wrinkles, or if someone or something wrinkles it, it gets folds or lines in it. （使）（衣服）起皱；起褶

absorbent [əbˈsɔːbənt] *adj.* Absorbent material soaks up liquid easily. 易吸收（液体）的；吸水的

inelastic [ˌɪnɪˈlæstɪk] *adj.* Something that is inelastic is unable to stretch easily and then return to its original size and shape. 有弹性的；有弹力的

breathable [ˈbriːðəbl] *adj.* A breathable fabric allows air to pass through it easily, so that clothing made from it does not become too warm or uncomfortable. （布料或织物）通气的；透气的

luster [ˈlʌstə] *n.* brightness; radiance; brilliance 光泽；光彩

drapability [ˌdræpəˈbɪlɪtɪ] *n.* the capacity to be draped 悬垂性

ornament [ˈɔːnəmənt] *n.* An ornament is an attractive object that you display in your home or in your garden. （家中或花园里的）装饰物；装饰品；点缀品

signify [ˈsɪɡnɪfaɪ] *v.* If an event, a sign, or a symbol signifies something, it is a sign of that thing or represents that thing. 表示；意味着；意思是

diplomatic [ˌdɪpləˈmætɪk] *adj.* Diplomatic means relating to diplomacy and diplomats. 外交的；外交官的

connotation [ˌkɒnəˈteɪʃn] *n.* The connotations of a particular word or name are the ideas or qualities which it makes you think of. （词或名字的）内涵意义；隐含意义；联想意义

stimulate [ˈstɪmjuleɪt] *v.* If you are stimulated by something, it makes you feel full of ideas and enthusiasm. 激发；激励；使充满热情

Phrases and expressions

account for （在数量、比例上）占
be honored as 被授予；被誉为
from afar 从远处；从远方；自远方；远道
eat one's words 收回前言；食言
wrap up 包裹起来
transform into 把…转变成…
pave the way for 为…铺平道路；为…做好准备

Chapter 7 Silk and Ceramics Culture

Proper names ...

Mawangdui Han Tomb　马王堆汉墓
Persia　波斯（西南亚国家，现在的伊朗）
Arabia　阿拉伯
Mediterranean　地中海的
Java　爪哇
Indian Ocean　印度洋
Red Sea　红海
Compendium of Materia Medica　《本草纲目》，明代李时珍著
Nongzhen Quanshu　《农政全书》，明代徐光启著，陈子龙修订
Tiangong Kaiwu　《天工开物》，明代宋应星著
Jianghai　江海，今江苏松江
Zhehai　浙海，今浙江宁波
Minhai　闽海，今福建泉州
Yuehai　粤海，今广东广州
Hangzhou Silkworm Academy　杭州蚕学馆

Cultural Notes ...

1. Hemudu culture（5000 to 4500 BC）was a Neolithic culture that flourished just south of the Hangzhou Bay in Jiangnan in modern Yuyao，Zhejiang，China. The culture may be divided into an early and late phases，before and after 4000 BC respectively. The site at Hemudu，22 km northwest of Ningbo，was discovered in 1973. Hemudu sites were also discovered on the islands of Zhoushan. Hemudu are said to have differed physically from inhabitants of the Yellow River sites to the north. Scholars view the Hemudu Culture as a source of the proto-Austronesian cultures.

Exercises ...

1. Read the following statements and try to decide whether it is true or false according to your understanding.

____ 1) Women, responsible for tilling to keep the family free of starvation, have played an important role in supporting family over thousands of years.

____ 2) In Shang Dynasty, silk products were only reserved for the royals and nobles for their own use or gifts to others.

____ 3) Figured silk, indicating that the weaving technique at that time reached a high level, emerged in the Warring States Period.

____ 4) Silk was once used as the materials for writing and painting.

____ 5) Silk is strong and elastic. If it is stretched, it can return to the same length immediately.

2. Fill in the blanks with the information you learn from the text.

1) In the small-scale peasant economy, men were supposed to till the land while women _____ at home.

2) Leizu tried and found that the fiber of the silkworm was strong and _____ enough to be used for cloth making.

3) In Qing Dynasty, the silk industry fell into a dilemma under exorbitant taxes and levies, and heavy _____ of foreign silk.

4) _____ embroidery is an excellent representative of Su embroidery.

5) China's silk has spread all over the world through _____. For example, silk goods starting from today's Xi'an City went westward to India, Persia, Arabia and finally to Mediterranean countries overland.

3. Explain the following terms.

1) Su Embroidery
2) Matouniang

4. Translate the following paragraph into English.

中国是丝绸之乡。在商代，织造技术达到了较高水平，出现了复杂的织机和提花织丝物。从西汉起，丝绸从长安出发经丝绸之路不断大批地运往国外。丝绸柔软、轻便、透气，却容易褪色、起皱，不耐脏。因其非凡的品质、精湛的工艺、精美的设计和色彩，丝绸受到普遍欢迎。它反映了中国古代工匠的集体智慧和对美的不懈追求。

5. Critical thinking and discussion.

With the improvement of craftsmanship and the increase of silk production, silk is no longer a tribute to the privileged. It is also accessible to ordinary people. But at present, silk is mostly used for making high-end clothing, ornament or gifts. As one of the oldest clothing materials in China, is silk going to decline and perish? Is there any possibility of new development in silk? If there is, in what fields will it take place?

Section B Culture of Ceramics

Chinese **ceramics**, including **pottery** and **porcelain**, are **renowned** throughout the world for their long history, great variety, artistic value and brilliant technique. The age-old ceramic wares reflect not only advances in technology but also improvements in the material and spiritual life of Chinese people. They constitute an important part of Chinese historical and cultural **heritage**.

Pottery Development

The invention of pottery marked the beginning of a new era in which mankind started to **exploit** and transform nature. The various kinds of pottery, ranging from household **utensils** to building materials and artistic **figurines**, give people today an insight into ancient societies.

Pottery in Prehistoric Period. Pottery making has a very long history in China. The earliest pottery in the world, found at the Xianrendong Cave Site （仙人洞遗址）in Jiangxi province, dates back to 20,000 years ago. And in the early New Stone Age, a lot more pottery **containers**, such as basins, jars and bowls, were produced. They were decorated with patterns like plants, animals, and **geometric** figures, which were of high artistic values. The pottery ware of Basin with a fish pattern with a human face (人面鱼纹彩陶盆) is a typical example, which was discovered in Banpo Village, Xi'an, Shaanxi Province in 1955. It is made from fine red clay. It has patterns of two opposing human faces and two fishes inside the basin. Both the mouth and the ears on the face are **depicted** as small fish, creating a peculiar merman-like form. On its head is a triangular-shaped object, possibly a **depiction** of hair, but also with the appearance of the fin of a fish. It reflects that fishing is one important means of life at that time.

Basin with a Fish Pattern with a Human Face

Pottery in Qin Dynasty. Pottery making industry went further in Qin Dynasty. The bricks and tiles used for the construction of the **Epang Palace**[1] had reached a high level. But what truly shocks the world is the **Terra Cotta Army** discovered near the tomb of Qin Shihuang, the first emperor unifying China in Chinese history. The army is composed of several thousand lifelike and life-sized terra cotta soldiers, terra cotta horses and bronze **chariots** and weapons, vividly representing the Qin military power and its powerful emperor. It is such a world-famous wonder that the site of Qin Shihuang Tomb was rated as **UNESCO World Heritage Site** in 1987. With their superb artistic creativity, Qin **craftsmen** left later generations with precious heritage of pottery and sculpture art.

Pottery in Han Dynasty. In Han Dynasty, there was greater progress in selecting and mixing clay to create **gigantic** food and water storage vessels. Besides, the craftsmanship of pottery figurine making reached a new height. Their decorative patterns are painted in bright colors, both pleasing to the eye and showing considerable artistic value. In contrast with the life-sized figurines created in a realistic way in Qin Dynasty, the figurines in this period became much smaller and were made in a much bolder and freer way. A typical representative is The pottery of Tomb Figurine of a Storyteller (击鼓说唱俑), which brought to life a performing artist 2,000 years ago with humorous facial expressions and exaggerated body postures. This figurine with its rich flavor of life tells us that storytelling became a popular form of entertainment at that time. The subject matters of pottery figures became more **diverse.** Except for the statues of music performers, soldiers, animals, some architectures and farmland scenes were also **modeled** into pottery. In a word, the pottery in Han Dynasty achieved a rather high artistic level.

Tomb Figurine of a Storyteller

Pottery in Tang Dynasty. In Tang Dynasty there was enormous variation in the colors of pottery wares. The most representative pottery was *Tang Sancai*, or tri-colored glazed pottery, which was invented on the basis of lead-glaze pottery of Han Dynasty. Its glaze mainly featured the three colors of yellow, green and white, hence the name. But *sancai* does not only refer to the three colors. Other basic glaze colors of blue, auburn（赤褐色）and black were also applied in an **interlaced** way. *Sancai* pottery was made not only into such traditional forms as bowls and vases, but also into the exotic fashion of camels and Central Asian travelers, testifying to the cultural influence of the Silk Road. The precious piece, A Troupe of Musicians on a Camel, provides us with a glimpse of the super techniques of *Tang Sancai* pottery and cultural **fusion** between China and Western Regions.

Tang Sancai pottery reached its **peak** in the prosperous period of Tang Dynasty but fell off in the late Tang Dynasty. Even though the glory was short-lived, its impact on the development of colored pottery and porcelain was great. *Tang Sancai* remains a most favored handicraft works for people today.

Porcelain Development

Ancient Chinese discovered and mastered porcelain making techniques for the first time in world history. Almost all peoples in the world had some sort of pottery making history, but the **feat** of porcelain creating was not achieved by all. This is because that porcelain must be manufactured at high temperatures in a complicatedly designed **kiln.** Besides, porcelain making also has strict requirements for raw material selection and **glaze** application. Porcelain is so **identified** with China that it is still called "china" in everyday English usage.

Porcelain in Han Dynasty. According to historical records, Chinese porcelain making started in the Eastern Han Dynasty. The basic material for porcelain making is a fine white clay called Kaolin（高岭土）, or china clay. In Shang Dynasty, people happened to make potteries with the clay. Potteries made of the material were smooth and **impermeable**, taking on the basic characteristics of porcelain. They were called primitive **celadon** by experts. In the Western Han Dynasty, along with the improvement of firing and molding techniques, the real celadon was created, which marked the beginning of porcelain making. The porcelain wares were produced mainly in Shangyu（上虞）area in Zhejiang Province. Among many kilns in this area, the Dragon Kiln（龙窑）was the most famous. It reached a mature firing temperature of over 1,200℃. The kiln was designed to increase or decrease temperature rapidly and to reach and maintain a **reduction atmosphere** as well. The porcelain produced in the kiln had glaze of a light green color and the body was white and **refined**. Since the Eastern Han Dynasty, the history of porcelain had opened a brand new **chapter**.

A Troupe of Musicians on a Camel

Porcelain in Tang Dynasty. Porcelain industry in Tang Dynasty reached its mature stage with the firing temperature kept between 1,250℃ and 1,450℃ and the products being very hard and thin. At this stage, most of the porcelain wares were single-colored, and the most famous among them were celadon and white porcelain. The celadon produced by Yue Kiln（越窑）, located in Yueyao County of Zhejiang Province, enjoyed a high reputation. In the book *Cha Jing*（《茶经》）, written by Tea Sage Lu Yu, tea sets produced at the Yue Kiln were regarded as the best. Poets of the Tang Dynasty spoke highly of the porcelain wares produced by the Yue kiln, saying that they were like **"jade"** and "ice" and their glaze looked like "layer upon layer of green peaks." White porcelain produced in the Xing Kiln (邢窑), located in Neiqiu County of Hebei Province, was "as white as snow or silver". In the late Tang Dynasty, the kilns in Changsha broke with the tradition of single-color glaze and started using **underglazed**-color (釉下彩) to produce multi-color porcelains.

Porcelain vessels were treated as precious and mysterious in the Muslim countries. This can be glimpsed from what Suleiman, an Arabian traveler and merchant, wrote in 851, "They have in China a very fine clay with which they make vases which are as transparent as glass; water is seen through them." In the late Tang Dynasty, lots of porcelain wares were exported to the Muslim countries along the Silk Road.

Porcelain in Song Dynasty. Porcelain industry flourished in Song Dynasty. Newly established kilns sprang up from all parts of the country. The most famous were kilns of Ru (汝), Guan (官), Ge (哥), Jun (钧) and Ding (定). All of them produced high quality porcelains, which were all imitated successfully by porcelain artisans in the following dynasties. Later, they were collectively known as "Five Great Kilns". Also, the kilns of

Five-footed Washer, Ge Ware

Yaozhou (耀州) in the north, and Longquan (龙泉) and Jingdezhen (景德镇) in the south contributed a lot in porcelain making. In this period, the techniques of porcelain making improved greatly, not only the single-colored porcelain, like celadon, white porcelain and black porcelain were produced, the multi-colored porcelain with **overglazing** or underglazing techniques were also started to make. Generally speaking, the methods in modeling, decorating, shaping and glazing were all innovated.

The porcelain in Song Dynasty distinguished itself from that in other Dynasties by its simple and quiet style. This style was rooted in the **aesthetic** idea in Song Dynasty to express natural order and find rhythm from within stillness and silence. Therefore, instead of using complicated patterned flowers for decoration, the vessels themselves might come in the shape of flowers. Another feature of Song porcelain is the creative use of crackle glaze as a form of decoration. The crack on the porcelain would appear when the glaze **contracted** unevenly in firing. By filling the crack with glazes, this technique flaw was transformed into a unique method of decoration. The "crab claw veins" (蟹爪纹) in Ru wares, "gold and iron threads (金丝铁线)" in Ge wares and "earthworm crawling in the mud (蚯蚓走泥纹)" in Jun wares are all examples of this technique.

With the rise of Islam (伊斯兰教) in Song Dynasty, porcelain vessels were in even higher demand in Muslim countries. At the same time, some decorative patterns with Islamic style made their way into China.

Jar with Underglaze Blue Design of Interlaced Peonies

Porcelain in Yuan Dynasty. In Yuan Dynasty, Chinese people came to regard white as a lucky color. This led to the popularity of qinghua porcelain or blue-and-white porcelain, a kind of underglaze colored porcelain, featuring pure, white paste and bright blue designs. In making such kind of porcelain, natural cobalt minerals served as coloring agent and Chinese calligraphy brush as the tool to paint the images and motifs. After the decorative patterns were drawn, the whole clay body would be covered with a transparent glaze and fired under high temperatures. The technique got its **debut** in Tang Dynasty and attained a level of maturity in Yuan Dynasty. Even though qinghua porcelains were produced in kilns from different places, none of them were the equal of those produced in Jingdezhen. It soon became an important porcelain center in China. Another wonderful achievement in this period was the creation of underglaze red porcelain (釉里红), which was created with the same technique but different coloring agent. The significance of the creation lies in that it started a porcelain making era featured by bright color decorations.

Porcelain in Ming Dynasty. China's porcelain industry reached the height of its prosperity in Ming Dynasty. Qinghua porcelain making remained the main stream in porcelain industry. In this period, qinghua porcelain gained a world-renowned reputation for its refined glazes and beautiful patterns that look like Chinese traditional **ink paintings.** Meanwhile, based on the making of blue-and-white porcelain, new techniques were developed. Different kinds of multi-color porcelain wares were produced. The most representative types were *wucai*（五彩） or five-color porcelain and *doucai*（斗彩） or contending-color porcelain.

Jar with *Wucai* Design of Coiled Dragon

Wucai wares in reality contains mostly three colors, that is red, green and yellow, within outlines of underglaze blue, plus the white of the porcelain body, all in all making up five colors. As time went on, more colors were used in *wucai* porcelain, but it kept the original name. *Doucai* porcelain was made by applying multiple layers of colors on blue-and-white porcelain vessels, which **"contended** for beauty". And this is the reason why it is called *doucai*. Both *wucai* and *doucai* are a combination of underglaze blue and overglaze colors. The difference lies in where the soft underglaze blue of *doucai* was primarily used for elegant outlines, the dark blue of *wucai* was applied.

Besides supplying porcelain for domestic use, the kilns in Ming Dynasty exported porcelain to the world on an **unprecedented** scale. Jingdezhen became the main production center for large-scale porcelain exports to Europe since the **reign** of Wanli Emperor.

Porcelain in Qing Dynasty. Qing Dynasty saw the peak of Chinese ceramic production in almost all ceramic types, especially during the reigns of Kangxi, Yongzheng, and Qianlong. It is a period especially noted for the production of color glazes. Qing potters succeeded in reproducing most of the famous glaze colors found in ceramic wares of the Song, Yuan and Ming dynasties. In addition, they adopted a variety of new glazes, thus bringing **vibrant** energy to Chinese porcelain art.

Famille-rose **enamels**（粉彩瓷釉）, "the pink family", were one creative addition to ceramic glazes during the Qing period. Porcelains with the enamel were known as *fencai* (粉彩). The enamels were first introduced from Europe around the end of the 1720s for use on porcelain. The majority of the enamels are **opaque** or semi-opaque and do not flow when fired. Compared with those transparent glazes used in *wucai* porcelain wares, these new opaque enamels could be fired at a lower temperature and had a wider color range. The colors also appeared softer and gentler, hence another name *ruan cai* (软彩). They are also known as *yangcai* (洋彩) since the enamels were originally imported from Europe. Technically, *Fencai* wares matured in the reign of Yongzheng and perfected in the reign of Qianlong. They are renowned as representing the peak of Qing porcelain manufacture as well as showing the fusion of Chinese and Western culture in the 18th century.

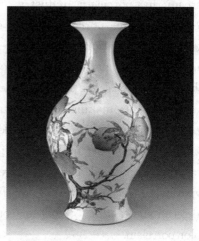

Vase with Famille Rose Design of Peaches and Bats

The quality of Chinese porcelain began to decline from the late Qing Dynasty as political instability took its toll on the arts. Fortunately, after the founding of People's Republic of China, the production of porcelain has **revived** and gained greater recognition at home and abroad.

Porcelain Capital of China

Jingdezhen is the world-famous Porcelain Capital. It was once called Xinping and Changnan in history. During the Jingde Period of the Southern Song Dynasty, Emperor Jingde **decreed** all the pieces made for court to be marked "made in the Jingde period" and hence the name Jingdezhen.

Historical records show that porcelain making in Jingdezhen dates back to Han Dynasty. Its most productive period, however, began from Song Dynasty and especially in the late Southern Song Dynasty when many ceramics workers from the north arrived in the city due to war. It was at the time that large quantities of porcelain wares were exported from Jingdezhen. In Yuan Dynasty the government office Fuliang Porcelain Bureau was set up, which attracted lots of skilled craftsman to the place and sped up the growth of porcelain industry. In Ming and Qing Dynasties, the ceramics industry in Jingdezhen flourished, producing high-quality wares on a large scale for export and for the imperial household as well. The porcelain produced there was so **exquisite** that it was as white as jade, as bright as a mirror, as thin as paper, with a sound as clear as a bell.

Today, Jingdezhen remains a national center for porcelain production. Every year many tourists come here for a visit to learn about porcelain culture.

Porcelain constitutes an essential part of Chinese people's life. It has been used by people in all sectors of society for both practical use and artistic appreciation. The superb quality of Chinese porcelain has also attracted foreigners and led to a flourishing export trade, which in turn serves as a bridge connecting China with other regions and countries in the world.

New words ...

ceramic [sɪˈræmɪk] n. Ceramic is clay that has been heated to a very high temperature so that it becomes hard. 陶瓷；陶瓷制品

pottery [ˈpɒtəri] n. You can use pottery to refer to pots, dishes, and other objects made from clay and then baked in an oven until they are hard. 陶器

porcelain [ˈpɔːsəlɪn] n. Porcelain refers to the delicate cups, plates, and ornaments made of made by heating clay. 瓷器

renowned [rɪˈnaʊnd] adj. A person or place that is renowned for something, usually something good is well known because of it. 有名望的；有声誉的

heritage [ˈherɪtɪdʒ] n. A country's heritage is all the qualities, traditions, or features of life there that have continued over many years and have been passed on from one generation to another. 遗产；传统

exploit [ɪk'splɔɪt] *v.* If you say that someone is exploiting you, you think that they are treating you unfairly by using your work or ideas and giving you very little in return. 利用；剥削

utensil [juː'tensəl] *n.* Utensils are tools or objects that you use in order to help you to cook, serve food, or eat. 器皿；烹调用具

figurine ['fɪɡəriːn] *n.* A figurine is a small ornamental model of a person. 小塑像

container [kən'teɪnə] *n.* A container is something such as a box or bottle that is used to hold or store things in. 容器

geometric [ˌdʒɪə'metrɪk] *adj.* Geometric or geometrical patterns or shapes consist of regular shapes or lines. 几何图形的

depict [dɪ'pɪkt] *v.* To depict someone or something means to show or represent them in a work of art such as a drawing or painting. 描绘

depiction [dɪ'pɪkʃn] *n.* A depiction of something is a picture or a written description of it. 描画；描述

chariot ['tʃærɪət] *n.* In ancient times, chariots were fast-moving vehicles with two wheels that were pulled by horses. 双轮马车

craftsmen ['krɑːftsmən] *n.* A craftsman is a man who makes things skilfully with his hands. 手艺人；工匠

gigantic [dʒaɪ'ɡæntɪk] *adj.* If you describe something as gigantic, you are emphasizing that it is extremely large in size, amount, or degree. 巨大的

diverse [daɪ'vɜːs] *adj.* Diverse people or things are very different from each other. 不同的

model ['mɒdl] *v.* I If you model shapes or figures, you make them out of a substance such as clay or wood. （用黏土或木头等）塑造

interlace [ɪntə'leɪs] *v.* To interlace something means to join together (patterns, fingers, etc) by crossing, as if woven. 使交织

fusion ['fjuːʒ(ə)n] *n.* A fusion of different qualities, ideas, or things is something new that is created by joining them together. 融合；熔合

peak [piːk] *n.* The peak of a process or an activity is the point at which it is at its strongest, most successful, or most fully developed. （过程、活动的）顶峰

feat [fiːt] *n.* If you refer to an action, or the result of an action, as a feat, you admire it because it is an impressive and difficult achievement. 功绩

kiln [kɪln] *n.* A kiln is an oven that is used to bake pottery and bricks in order to make them hard. 窑

glaze [ɡleɪz] *n. v.* A glaze is a thin layer of liquid which is put on a piece of pottery and becomes hard and shiny when the pottery is heated in a very hot oven. 釉

To glaze means to cover plates, cups etc. made of clay with a thin liquid that gives them a shiny surface. 给…上釉

identify [aɪˈdentɪˌfaɪ] *v.* If you identify one person or thing with another, you think that they are closely associated or involved in some way. 认为…密切相关

impermeable [ɪmˈpɜːmɪəbəl] *adj.* Something that is impermeable will not allow fluid to pass through it. 无法渗透的

celadon [ˈselədɒn] *n.* Celadon is a type of porcelain having a greyish-green glaze. 灰绿色瓷器

refine [rɪˈfaɪn] *v.* When a substance is refined, it is made pure by having all other substances removed from it. 提炼

chapter [ˈtʃæptə] *n.* A chapter in someone's life or in history is a period of time during which a major event or related events takes place. （事件发生的）时期

jade [dʒeɪd] *n.* Jade is a hard stone, usually green in color that is used for making jewelry and ornaments. 玉

underglaze [ˈʌndəɡleɪz] *v.* To underglaze means to apply colors or decorations to porcelain or pottery before the application of glaze 上釉下彩

overglaze [ˈəʊvəɡleɪz] *v.* To overglaze means to apply colors or decorations to porcelain or pottery above the glaze 给…施第二层釉

aesthetic [iːsˈθetɪk] *adj.* Aesthetic is used to talk about beauty or art, and people's appreciation of beautiful things. 审美的

contract [kənˈtrækʃ(ə)n] *v.* When something contracts or when something contracts it, it becomes smaller or shorter. 缩小；缩短

debut [ˈdeɪbjuː] *n.* The debut of something new or something important refers to their first public appearance. 新事物的问世

contend [kənˈtend] *v.* If you contend with someone for something such as power, you compete with them to try to get it. 争夺权力等

unprecedented [ʌnˈpresɪdentɪd] *adj.* If you describe something as unprecedented, you are emphasizing that it is very great in quality, amount, or scale. 质量、数量或规模空前的

reign [reɪn] *n.* The period when a king or queen rules a country 统治时期

vibrant [ˈvaɪbrənt] *adj.* Someone or something that is vibrant is full of life, energy, and enthusiasm. 充满活力的

opaque [əʊˈpeɪk] *adj.* If an object or substance is opaque, you cannot see through it. 不透明的

revive [rɪˈvaɪv] *adj.* When something such as the economy, a business, a trend, or a feeling is revived or when it revives, it becomes active, popular, or successful again. 恢复；复兴

decree [dɪˈkriː] *v.* If someone in authority decrees that something must happen, they decide or state this officially. 发布命令

exquisite [ˈekskwɪzɪt] *adj.* Something that is extremely beautiful or pleasant, especially in a delicate way. 精美的

Phrases and expressions

range from sth to sth　包括
date back to　追溯到，从…就开始有
be composed of　由某物组成

be rated as 被评为
testify to 证明，证实
be pleasing to the eye 悦目，美观
fall off 逐渐消失，消亡，死亡，废弃
identify sth with sth 将某物等同于另一物；认为某物与另一物有密切关联
break with 放弃，摒弃
spring up 涌现
be rooted in 起源于某事物
put an end to 结束，终止
on an unprecedented scale 以史上空前的规模
take a toll on 严重损害

Proper names

Terra Cotta Army 兵马俑，即秦始皇兵马俑，亦简称秦兵马俑或秦俑，第一批全国重点文物保护单位，第一批中国世界遗产，位于今陕西省西安市临潼区秦始皇陵以东1.5千米处的兵马俑坑内。

UNESCO World Heritage Site 世界文化遗产，是一项由联合国发起、联合国教育科学文化组织负责执行的国际公约建制，以保存对全世界人类都具有杰出普遍性价值的自然或文化处所为目的。世界文化遗产是文化的保护与传承的最高等级，属于世界遗产范畴。

Reduction Atmosphere 还原气氛，指在陶瓷烧制过程中温度达到1 100～1 300℃高温时人为地使窑里面充满一氧化碳等还原性气体而含氧量极低的氛围。

Ink Painting 水墨画是指经过调配水和墨的浓度所画出的画，是绘画的一种形式，水墨画被视为中国传统绘画，也称国画。

enamel 珐琅，用石英、长石、硝石和碳酸钠等加上铅和锡的氧化物烧制成的像釉子的物质。用它涂在铜质或银质器物上，经过烧制，形成不同颜色的釉质表面，既可防锈，又可作为装饰。如搪瓷、景泰蓝等均为珐琅制品。

Cultural Notes

1. Epang Palace was built in Qin Dynasty in 212 BC, with the site located in E-Pang Village, western suburb of Xi'an City, Shaanxi Province. After Qin Shi Huang united China, the national power was undergoing continuous growth. In 212 BC, Qin Shi Huang commanded to build E-Pang Palace. However, due to the huge scale of the project, only a front palace was completed during his reign. According to historical records, the front palace of E-Pang covers an area about 80,000 square meters. In 1994, relics of E-Pang Palace was confirmed by the UNESCO as one of the miracles and famous relics in the world and crowned as "The First Palace under Heaven".

Exercises

1. Read the following statements and try to decide whether it is true or false according to your understanding.

_____ 1) Although porcelain is developed from pottery, the two are different in raw material, glaze and firing temperature.

_____ 2) Porcelain making started in the Eastern Han Dynasty when the kilns reached a firing temperature over 1,200℃.

_____ 3) Porcelain industry in Tang Dynasty entered its mature stage and it is famous for the celadon in the South and the white porcelain in the North.

_____ 4) The fashionable porcelain wares in Song Dynasty looked simple and quiet without complicated decorations.

_____ 5) It was in Yuan Dynasty the kilns exported porcelain around the world on an unprecedented scale.

2. Fill in the blanks with the information you learn from the text.

1) _____ is a collection of several thousand sculptures of soldiers, horses, chariots and weapons, depicting the armies of the first emperor of China.

2) The most famous multi-color porcelains in Ming Dynasty are _____ and _____ .

3) _____, _____, _____, _____ and _____ were often referred to as "five great kilns" in Song Dynasty because the porcelains produced there were of high quality.

4) Blue and white porcelain, features pure, white paste and bright blue designs, was first produced in _____ Dynasty.

5) During the Qing Dynasty, multi-color porcelains were enriched with the _____ porcelain.

3. Explain the following terms.

1) *Tang Sancai*
2) The Porcelain Capital of China

4. Translate the following paragraph into English.

瓷器是一种由精选的瓷土或瓷石，通过工艺流程制成的物件。尽管瓷器由陶器发展而来，然而它们在原料、釉、烧制温度方面都是不同的。比起陶器，瓷器质地更坚硬，器身更通透，光泽也更好。瓷器促进了中国和外部世界之间的经济文化交流，并深刻影响着其他国家的传统文化和生活方式。

5. Critical thinking and discussion.

Auctions of cultural relics overseas have always been a controversial issue in China. Many historic artifacts were looted during wars and other periods of unrest in China from the late 19th century to early 20th century. The number of items stolen is estimated to be around 10 million, including from museums and private collections, according to the China Cultural Relics Academy. The State Administration of Cultural Heritage issued a regulation in 2016 to ban domestic auctions of Chinese cultural relics that had been previously looted illegally. But it is hard to bring these relics back, as museums refuse to return the relics, while sellers and buyers in auctions are often anonymous. Once the deal is made, it is difficult to trace the relics. Nevertheless, some patriotic Chinese collectors have bought the lost relics at auctions and donated them to the Chinese government in the past few years.

1) How do you think about these patriotic actions to bring lost relics back

home?

2) Do you have better ways to bring lost relics back home?

Section C The Silk Road

The Silk Road was an ancient network of trade routes that were for centuries central to cultural interaction through many regions of **Eurasia**. It stretched from China to the Korean peninsula and Japan in the east, and connected China through Central Asia to India in the south and to the Mediterranean Sea in the west. It was formally established during Han Dynasty, which linked many regions of Asian and European countries in commerce. Until now, the Silk Road System has existed for over 2,000 years, with specific routes changing over time. It was a network consisting of both land and sea trading routes, with the Silk Road on land being much more influential and **impressive**. Therefore, the name Silk Road can still **evoke** images of trade caravans **wading** through desert sand and smell of **exotic** spices, and continue to fascinate travelers. The long history of the Road, closely linked with China, explains the political and economic reasons for its success.

The Term "Silk Road"

For thousands of years, along the Road traveled highly valued silk, cotton, wool, glass, jade, gold, silver, salt, tea, herbal medicines, foods, fruits and even religious ideas. Among these goods, silk comprised a large proportion of trade along this road. So a German geographer and traveler, Ferdinand von Richthofen, **coined** two names for this network of trade routes in 1877. One was called Seidenstrassen (Silk Routes) and the second Seidenstrasse (Silk Road). The former was favored by historians, because this network was not a single **thoroughfare** from east to west. While the latter has increasingly become a more common and **recognized** name. Although Silk Road derived its name from the main traded goods—silk along the road, it gave people not only the chance to exchange goods but cultures and ideas between China and the West.

Silk Road in Han Dynasty

This route was opened up by **Zhang Qian**[1] in the Western Han Dynasty. At that time, the northern and western borders were regularly **harassed** by the **nomadic** tribes of Xiongnu. The caravans were often attacked by some small Central Asian tribes. In order to assure the safety of the trade, the Han Government sent General Zhang Qian as an **envoy** to build relationships with these small nomadic tribes in 138 BC. Starting from Chang'an, the capital of the Western Han Dynasty, and crossing the vast Western Regions, Zhang reached

Zhang Qian

Loulan, **Qiuzi**, and **Yutian** and established trade relations with these small but important kingdoms. Zhang's officers went even further into Central Asia. All of the kingdoms Zhang and his **delegation** visited sent their envoys to Chang'an to express their appreciation for the new relationship and show their respect to the Han government. From then on, the merchants traveled on this route safely and began to carry silk from China to other parts of the world.

After Zhang Qian, the Eastern Han General and **diplomat Ban Chao**[2] led an army in 73 to secure the trade routes, reaching far west to the Tarim Basin （塔里木盆地） in Xinjiang Autonomous region. Ban Chao expanded his conquests across the Pamirs （帕米尔高原） in the most western end of China to the Caspian Sea （里海） and Iran-based **Parthia**. Ban Chao lived in the Western Regions for about 31 years, put down **countless** rebellions and built diplomatic relations with more than 50 states, ensuring peace and stability along the Silk Road. In 97, Ban Chao **dispatched** envoy **Gan Ying**[3] to Daqin （the Roman Empire） for trading with Romans directly. He began his journey from Qiuzi and finally reached the Persian Gulf. Being afraid of crossing the gulf, he gave up halfway. Although he didn't finish his **mission**, he brought more reliable information about Central Asia. In 123, Ban Yong （班勇）, the son of Ban Chao, was appointed to protect this trade road and he defeated a king of Xiongnu. Ban Yong **consolidated** the reign of the Eastern Han Dynasty in the West Regions and greatly protected the ancient trade road.

In a word, the Silk Road gradually came into being in Han Dynasty. Since then, China made a lot of efforts to consolidate the road to the Western Regions and India, such as encouraging Chinese settlements in the area of the Tarim Basin and establishing the diplomatic relations with some countries in the Western Regions such as **Dayuan**.

Silk Road in Tang Dynasty

Although the Silk Road was initially **formulated** during the **reign** of **Emperor Wu of Han**[4] in the Western Han Dynasty, the Road rose to its most **flourishing** period in Tang Dynasty. Its Golden Age of development can be attributed to three reasons. First, the economy developed very rapidly, which helped Tang to be very powerful and **prosperous** in Chinese history. This allowed it to develop an **unprecedented** trade with the outside world. Second, the Chinese emperors welcomed foreign cultures, making Tang very **cosmopolitan** in its urban centers. At that time, the capital Chang'an (now Xi'an) was an international **metropolis** in the world and it attracted thousands of traders and tourists from around the world, especially Persians and **Sogdians**. Dunhuang in **Hexi Corridor**, as the gate to Western Regions, was also a significant city and attracted a lot of foreigners. Chinese people also traveled along the Silk Road to visit other countries. For example, the famous Chinese monk **Xuan Zang** traveled along this Road to India to explore the true nature of **Buddha**. Third, China built good relations with many countries in the west such as Persia and Arabian Empire. This created a relatively stable external environment for the trade prosperity along the Road.

In Tang Dynasty the overland Silk Road was greatly extended through the **Mongolian Plateau** in the north and to the Qinghai-Tibet Plateau in the south. In the north, Tang conquered **Gaochang**, **Yanqi**, Qiuzi of Western Regions and controlled Mongolian Plateau in the 7th century. This helped Tang to strengthen the trade relations and contacts with Western Regions and Mongolian Plateau. In the south, Tang built friendly relations with Tibet by the marriage between **Princess Wencheng** and **Sontzen Gampo** (the ruler of Tibet) in 641. Since then, the trade and cultural exchanges became more frequent between Tang and Tibet. Thus, the ancient Silk Road closely linked with the Qinghai-Tibet Plateau. Without doubt, the Silk Road flourished and reached its peak in Tang Dynasty. Unfortunately, after **Anshi Rebellions** Tang began to decline from prosperity, so did the Silk Road.

In addition to the land route, Tang also developed the **Maritime** Silk Road. Chinese envoys had been sailing through the Indian Ocean to India since about the 2nd century BC, yet it was till Anshi Rebellions that this route was regarded as a secondary **alternative** to the overland Silk Road. However, with the technological advances in shipbuilding and **navigation** the trade **boomed** along the Maritime Silk Road. The Maritime Silk Road was extended to the Southeast Asia, the Persian Gulf and Red Sea, even into Somalia in the Horn of Africa. At that time, Guangzhou became the first great trade harbor in China and it made a great contribution to the development of the maritime Silk Road.

Silk Road in Yuan Dynasty

Along with the establishment of Yuan Dynasty, the Silk Road regained its **vigor** and became prosperous once again. During Yuan Dynasty, the Silk Road experienced its last flourishing period.

In 1271, the great Mongolian ruler **Kublai Khan** established the powerful Yuan Dynasty at Dadu, the modern Beijing. Yuan destroyed a great number of toll-gates and **corruption** on the Silk Road; therefore it became more convenient, easier and safer than ever before to pass through the Silk Road. The Yuan emperors welcomed western travelers and even **appointed** some foreigners high positions. For example, Kublai Khan gave **Marco Polo** a **hospitable**

Marco Polo

welcome and appointed him a high post in his court. At that time, Yuan issued a special VIP passport known as "Golden Tablet" which entitled holders to receive food, horses and guides throughout its **dominion**. The holders were able to travel freely and carried out trade between East and West directly in the **realm** of Yuan Dynasty. It is thus clear that the opening-up policy of Yuan greatly contributed to the development of Silk Road in the north, connecting the Mongolia Plateau and the European region north of Mediterranean. This is well-known Grassland Silk Road.

Besides, the Maritime Silk Road boomed rapidly in the south of China during the period of Yuan mainly due to the stable environment and the fast economic development, especially the navigation technology. Thus the flourishing maritime trade had a large impact on the Road in the north. Despite all this, many westerners, Chinese envoys and caravans traveled along this ancient trade route on the land. However, it eventually could not **contend** with the expansion of shipping and thus started its **demise**.

Decline of Silk Road

The reasons for the decline of Silk Road could be attributed to three reasons. First, the **disintegration** of Yuan **loosened** the political, cultural, and economic control and protection of the Silk Road. Second, Ming and Qing Dynasties adopted the closed-door policy which **suspended** the foreign trade. Third, the ancient countries of Western Regions along the Silk Road gradually disappeared due to years of war and violence, which partly caused the close of Silk Road. Although the consolidation of **Ottoman Empire** and **Safavid Empire** in the Middle East led to a **revival** of overland trade, it was often interrupted by warfare between them. It was further disrupted by the **collapse** of Safavid Empire in the 1720s. The closing of the Silk Road on land forced merchants to take advantage of the sea to **ply** their trade, thus initiating the **Age of Discovery**[5] which led to the world-wide interaction and the beginnings of a global community.

Overall, the ancient Silk Road has four main routes connecting China and the West. The first is the Northwestern Silk Road, the official route opened by Zhang Qian in the Western Han Dynasty. It starts from Chang'an, running through the Central Asia, Afghanistan, Iran, Iraq and so on, and finally reaches the Mediterranean and Rome. This road is considered as the **intersection** of Eastern and Western civilizations, and it is the so-called Silk Road. The second is Grassland Silk Road, from Chang'an through Dunhuang, **Urumqi** and Central Asia to Rome. The third is the Southwest Silk Road from Xi'an via Chengdu to India, extending to South Asia, Central Asia, and European countries. The fourth is the Maritime Silk Road, from Guangzhou, Quanzhou, Hangzhou and other coastal cities to Arabian Sea, even to the east coast of Africa.

In general, the Silk Road is the road of **integration**, exchange and dialogue between East and West. It has made important contributions to the common prosperity of mankind in the past 2,000 years. Besides, the commercial goods featuring "silk" and "**embroidery**" as the main products, science and technology and every other element of civilization were also introduced into West, for example, the four great inventions of China, including papermaking, compass, gunpowder and printing. Undoubtedly, the Silk Road has promoted the economic and cultural exchanges between China and the West. Now it still remains an important channel for exchanges between Chinese and the West.

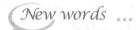

impressive [ɪmˈpresɪv] *adj.* (of things or people) making you feel admiration, because they are very large, good, skilful, etc. 事物或人令人赞叹的；令人敬佩的

evoke [ɪˈvəʊk] *v.* To evoke a particular memory, idea, emotion, or response means to cause it to occur. 唤起；召唤；引起

wade [weɪd] *v.* If you wade through something that makes it difficult to walk, usually water or mud, you walk through it. （艰难地）涉；蹚；跋涉

exotic [ɪɡˈzɒtɪk] *adj.* Something that is exotic is unusual and interesting, usually because it comes from or is related to a distant country. 具有异国情调的；外来的；奇异的

coin [kɔɪn] *v.* If you coin a word or a phrase, you are the first person to say it. 新造，杜撰（单词、短语）

thoroughfare [ˈθʌrəfeər] *n.* A thoroughfare is a main road in a town or city which usually has shops along it and a lot of traffic. 大道；大街

recognized [ˈrekəɡnaɪzd] *adj.* If someone says something is recognized, this means that they acknowledge that it exists or that it is true. 公认的

harass [ˈhærəs] *v.* If someone harasses you, they trouble or annoy you, for example by attacking you repeatedly or by causing you as many problems as they can. 骚扰；烦扰

nomadic [nəʊˈmædɪk] *adj.* Nomadic people travel from place to place rather than living in one place all the time. 游牧的；游牧部落的

envoy [ˈenvɔɪ] *n.* An envoy is someone who is sent as a representative from one government or political group to another. 使节；使者；代表

delegation [ˌdelɪˈɡeɪʃn] *n.* A delegation is a group of people who have been sent somewhere to have talks with other people on behalf of a larger group of people. 代表团

diplomat [ˈdɪpləmæt] *n.* A diplomat is a senior official who discusses affairs with another country on behalf of his or her own country, usually working as a member of an embassy. 外交官；外交家

countless [ˈkaʊntləs] *adj.* Countless means very many. 无数的；数不尽的

dispatch [dɪˈspætʃ] *v.* If you dispatch someone to a place, you send them there for a particular reason. 派遣；调遣

mission [ˈmɪʃn] *n.* An important official job that a person or group of people is given to do, especially when they are sent to another country. 官方使命；使团使命

consolidate [kənˈsɒlɪdeɪt] *v.* If you consolidate something that you have, for example power or success, you strengthen it so that it becomes more effective or secure. 加强；巩固

formulate [ˈfɔːmjuleɪt] *v.* If you formulate something such as a plan or proposal, you invent it. 规划；策划

reign [reɪn] *n.* When a king or queen reigns, he or she rules a country. （国王或女王）统治，治理

flourish [ˈflʌrɪʃ] *v.* If something flourishes, it is successful, active, and develops quickly and strongly. 繁荣

prosperous [ˈprɒspərəs] *adj.* Prosperous places and economies are rich and successful. 富足的；兴旺的；繁荣的

unprecedented [ʌnˈpresɪdentɪd] *adj.* If you describe something as unprecedented, you are emphasizing that it is very great in quality, amount, or scale. （数量、规模）很大的

cosmopolitan [ˌkɒzməˈpɒlɪtən] *adj.* A cosmopolitan place or society is full of people from many different countries and cultures. 世界性的；国际化的；受各国文化影响的

metropolis [məˈtrɒpəlɪs] *n.* A metropolis is the largest, busiest, and most important city in a country or region. 大都市

Buddha [ˈbʊdə] *n.* Buddha is the title given to Gautama Siddhartha, the religious teacher and founder of Buddhism. 佛陀（对佛教创始人乔答摩·悉达多的尊称）

alternative [ɔːlˈtɜːnətɪv] *n.* If one thing is an alternative to another, the first can be found, used, or done instead of the second. 可供选择的事；可供替代的事

navigation [ˌnævɪˈɡeɪʃn] *n.* You can refer to the movement of ships as navigation. 航行；航海

boom [buːm] *v.* If the economy or a business is booming, the amount of things being bought or sold is increasing. （经济或生意）景气，繁荣

vigor [ˈvɪɡə] *n.* Vigor is physical or mental energy and enthusiasm. 活力

corruption [kəˈrʌpʃn] *n.* Corruption is dishonesty and illegal behavior by people in positions of authority or power. 腐败；贪污；受贿

appoint [əˈpɔɪnt] *v.* If you appoint someone to a job or official position, you formally choose them for it. 任命；委派；指派

hospitable [hɒˈspɪtəbl] *adj.* A hospitable person is friendly, generous, and welcoming to guests or people they have just met. 热情的；好客的

dominion [dəˈmɪniən] *n.* A dominion is an area of land that is controlled by a ruler. 领土；版图；领地

realm [relm] *n.* A realm is a country that has a king or queen. 王国

contend [kənˈtend] *v.* If you contend with someone for something such as power, you complete with them to try to get it. 竞争；争夺

demise [dɪˈmaɪz] *n.* The demise of something or someone is their end or death. 终止；消亡；死亡

disintegration [dɪsˌɪntɪˈɡreɪʃn] *n.* The disintegration of something means that it becomes seriously weakened, and is divided or destroyed. 崩溃；瓦解

loosen [ˈluːsn] *v.* If a government or organization loosens its grip on a group of people or an activity, or if its grip loosens, it begins to have less control over it. 放宽；放松（控制）

suspend [səˈspend] *v.* If you suspend something, you delay it or stop it from happening for a while or until a decision is made about it. 暂缓；推迟；暂停

revival [rɪˈvaɪvl] *n.* When there is a revival of something, it becomes active or popular again. 复苏；复兴

collapse [kəˈlæps] *n.* It means a system or institution fails or comes to an end completely and suddenly. 瓦解；崩溃

ply [plaɪ] *v.* If you ply a trade, you do a particular kind of work regularly as your job, especially a kind of work that involves trying to sell goods or services to people outdoors. 从事；经营

intersection [ˌɪntəˈsekʃn] *n.* An intersection is a place where roads or other lines meet or cross. 交点；道路交叉口

integration [ˌɪntɪˈɡreɪʃn] *n.* It means the act of combining into an integral whole. 融入；融合

embroidery [ɪmˈbrɔɪdəri] *n.* Embroidery is the activity of stitching designs onto cloth. 刺绣

Chapter 7 Silk and Ceramics Culture

Phrases and expressions

wade through　涉（水，泥泞等）；费力前进；很吃力地通过
link...with...　将…与…连接在一起
put down　镇压
come into being　产生；诞生
attribute...to...　把…归因于…
make a great contribution to...　对…做出重大贡献；有利于…
contend with　与某事物抗争；苦于应付

Proper names

Eurasia　亚欧大陆，是亚洲大陆和欧洲大陆的合称，也称欧亚大陆，面积达 5 000 多万平方千米。
Loulan　楼兰国，是西域古国名，是中国西部的一个古代小国，国都楼兰城，遗址在今中国新疆罗布泊西北岸。
Qiuzi　龟兹国，又称丘慈、邱兹、丘兹等，是中国古代西域大国之一，汉朝时为西域北道诸国之一，唐代安西四镇之一。1758 年定名为库车，隶属于新疆维吾尔自治区。
Yutian　于阗（tian）国（公元前 232—1006）是古代西域王国，中国唐代安西都护府安西四镇之一。地处塔里木盆地南沿。
Parthia　帕提亚，亚洲西部古国，在伊朗东北部。
Dayuan　大宛国，古代中亚国名，大宛国大概在今费尔干纳盆地。
Sogdian　古索格代亚纳人，居住在索格代亚纳的伊朗人。
Hexi Corridor　河西走廊，古称雍州、凉州，简称"河西"，是中国内地通往西域的要道。位于甘肃西北部，为南北走向的长条堆积平原，因位于黄河以西，故名。
Xuan Zang　玄奘（602—664），唐代高僧，俗家姓名"陈祎（yi）"，法名"玄奘"。

Mongolian Plateau 蒙古高原泛指亚洲东北部高原地区，亦即东亚内陆高原，东起大兴安岭、西至阿尔泰山，北界为萨彦岭、雅布洛诺夫山脉，南界为阴山山脉，范围包括蒙古全境和中国华北北部，面积约200万平方公里。

Gaochang 高昌国是汉族在西域建立的佛教国家，位于今新疆吐鲁番市高昌区东南方向，是古时西域交通枢纽。

Yanqi 焉耆，又称乌夷、阿耆尼，新疆塔里木盆地古国，今在新疆维吾尔自治区焉耆回族自治县附近。

Princess Wencheng 文成公主原本是李唐远支宗室女，唐太宗贞观十四年（640），太宗李世民封李氏为文成公主；贞观十五年（641）文成公主远嫁吐蕃，成为吐蕃赞普松赞干布的王后。

Sontzen Gampo 吐蕃王朝立国之君，在位期间（629—650），平定吐蕃内乱，统一西藏，正式建立奴隶主统治的吐蕃王国。迎娶文成公主后，唐封他为驸马都尉、西海郡王。

Anshi Rebellions 安史之乱是中国唐代玄宗末年至代宗初年（755—763）由唐朝将领安禄山与史思明背叛唐朝后发动的战争，为唐由盛而衰的转折点。

Kublai Khan 忽必烈（1216—1294），元世祖，中国元朝皇帝，成吉思汗之孙。

Marco Polo 马可·波罗（1254—1324），威尼斯旅行家、商人，著有著名的《马可·波罗游记》。

Ottoman Empire 奥斯曼帝国（1299—1922），土耳其人建立的帝国，创立者为奥斯曼一世，地跨欧非亚的帝国。

Safavid Empire 萨非王朝（1502—1736），波斯人建立统治伊朗的王朝。

Urumqi 乌鲁木齐，是新疆维吾尔自治区首府。它是清朝在新疆驻军而发展起来的一座城市，1763年改称"迪化"，1884年为新疆省省会，从此成为全疆政治中心。1954年迪化改称"乌鲁木齐"。

Cultural Notes

1. Zhang Qian（张骞）was a Chinese official and diplomat who served as an imperial envoy to the world outside of China in the 2nd century BC, during the time of Han Dynasty. He was the first official diplomat to bring back reliable information about Central Asia to the Chinese imperial court. Under Emperor Wu of Han he played an important pioneering role in the Chinese colonization and conquest of the region now known as Xinjiang.

2. Ban Chao（班超）was a Chinese general, explorer and diplomat of the Eastern Han dynasty. He was born in Fufeng, now Xianyang, Shaanxi. As the Han general and cavalry commander, Ban Chao was in charge of administrating the Western Regions（Central Asia）while he was in service. He also led Han forces for over 30 years in the war against Xiongnu and secured Han control over the Tarim Basin region. He was awarded the title "Protector General of the Western Regions" by the Han government for his efforts in protecting and governing the regions.

3. Gan Ying（甘英）was a Chinese military ambassador who was sent on a mission to Rome in 97 by the Chinese general Ban Chao. Although Gan Ying never reached Rome, only travelling to as far as the "western sea" which either refers to the Black Sea or the Parthian coast of the Persian Gulf, he is, at least in the historical records, the Chinese who went the furthest west during the ancient time.

4. Emperor Wu of Han（刘彻）was the seventh emperor of Han Dynasty of China, ruling from 141-87 BC. It was also during his reign that cultural contact with western Eurasia was greatly increased, directly or indirectly. Many new crops and other items were introduced to China during his reign.

5. Age of Discovery was a period from the early 15th century that continued into the early 17th century, during which European ships traveled around the world to search for new trading routes and partners. They were in search of trading goods such as gold, silver and spices. In the process, Europeans met peoples and mapped lands previously unknown to them.

Exercises

1. Read the following statements and try to decide whether it is true or false according to your understanding.

____ 1) The Silk Road was an ancient network of trade routes connecting China to the West.

____ 2) The Silk Road was a trade road between China and the West in which the Chinese and western merchants did the silk trade.

____ 3) The Silk Road opened in Han dynasty and experienced its Golden Age of development in Yuan Dynasty.

____ 4) In Yuan Dynasty, the government cleared a great many toll-gates to welcome westerners to trade with Chinese people.

____ 5) The fragmentation of the Mongolian Empire directly caused the decline of Silk Road.

2. Fill in the blanks with the information you learn from the text.

1) Silk Road can be classified into the Silk Road on the land and the Silk Road _____ .

2) Han dynasty was regularly harassed by _____ of Xiongnu on the northern and western borders.

3) The Silk Roads were a complex network of trade routes that gave people the chance to exchange goods and _____ .

4) During Tang dynasty the Chinese monk _____ traveled along the Silk Road to India to explore the true nature of Buddha.

5) The Mongolian emperor issued a special _____ known as "Golden Tablet" which entitled holders to receive food, horses and guides throughout

its dominion.

3. Explain the following terms.

    1) Origin of Silk Road

    2) Golden Age of Silk Road

    3) Decline of Silk Road

4. Translate the following paragraph into English.

    闻名于世的丝绸之路是一系列连接东西方的路线。丝绸之路延伸6 000多千米，得名于古代中国的丝绸贸易。丝绸之路上的贸易在中国、南亚、欧洲和中东文明发展中发挥了重要作用。正是通过丝绸之路，中国的造纸、火药、指南针、印刷等四大发明才被引介到世界各地。同样，中国的丝绸、茶叶和瓷器也传遍全球。物质文化的交流是双向的。欧洲也通过丝绸之路出口各种商品，满足中国市场的需要。

5. Critical thinking and discussion.

    Chinese scholar, Ji Xianlin, once said: Even people with a little knowledge know the Silk Road across the Eurasia. It is in fact a large artery of cultural exchanges between East and West during a very long historical period. It widely and deeply influenced our China and the countries along the road in politics, economy, culture, art, religion, and philosophy. If there is no such a road, the development of these countries today and how exactly, we simply cannot imagine.

    1) How do you understand the meaning of the Silk Road according to the above comment?

    2) How do you comment on the value and implications of Silk Road for our country's economic development?

Section D The Belt and Road Initiative

    The Belt and Road **Initiative** is a global development strategy proposed by the Chinese government. It refers to the **Silk Road Economic Belt** and **21st Century Maritime Silk Road** and focuses on connectivity and cooperation between Eurasian countries along the Belt and Road routes. The Initiative is designed to enhance the

efficient allocation of resources, further market **integration** and establish a framework for regional economic cooperation for all concerned countries.

Historical Background

The world economy is facing various challenges. Some international financial problems continue to emerge. And the world economy is recovering slowly and developing unevenly. Under this circumstance, it is urgent to adhere to open spirit to develop regional cooperation for an inclusive and balanced cooperation framework as well as orderly free flow and **optimized** allocation of resources.

Historically, the overland and maritime silk roads have been the major routes for economic, trade and cultural exchanges between China and other countries in Asia, Africa, and Europe. In modern time, the Initiative will help **coordinate** the development strategies of the countries along the Belt and Road, **excavate** the potentials in the regional market, promote investment and consumption, and create job opportunities. It can also enhance cultural exchanges of the people along the routes, make them know, trust and respect each other, and share peace and prosperity.

Currently, China's economy is highly related to the world economy. In order to realize **comprehensive**, balanced and **sustainable** economic development, China must obey the basic policy of opening-up and integrate deeply into the world economy. The Belt and Road Initiative complies with the trend toward economic globalization. It helps to drive the countries concerned to coordinate their economic policies for a globally free trade system and an open world economy. Promoting the construction of Belt and Road is not only the need for China to expand and deepen opening-up, but also the need for the countries in Asia and Europe to strengthen mutually beneficial cooperation.

Conceptual Framework

In September 2013, during a state visit to Central Asia and Southeast Asia, President Xi Jinping raised the Initiative to build a Silk Road Economic

Belt. The Initiative calls for a **cohesive** economic area for all regions through building **infrastructure**, strengthening cultural exchanges, and expanding trade. In October 2013, the Maritime Silk Road Initiative was put forward by Xi Jinping in a speech to the Indonesian Parliament. The Maritime Silk Road is also called the "21st Century Maritime Silk Road". As a **complementary** part, it aims to invest and foster collaboration in Southeast Asia, Oceania, and North Africa through several neighboring bodies of water.

In short, the Initiative aims to connect Asia, Europe and Africa along five routes. The Silk Road Economic Belt focuses on: linking China with the Persian Gulf and the Mediterranean Sea through Central Asia and Russia; connecting China with the Middle East through Central Asia; and bringing together China and Southeast Asia, South Asia and the Indian Ocean. The Maritime Silk Road focuses on: linking China with Europe through the South China Sea and Indian Ocean; and connecting China with the South Pacific Ocean through the South China Sea. Based on the above five routes, six international economic cooperation **corridors** are jointly built. These have been identified as the **New Eurasia Land Bridge**, China-**Mongolia**-Russia, China-Central Asia-West Asia, **China-Indochina Peninsula**, China-**Pakistan**, and **Bangladesh**-China-India-**Myanmar**.

Cooperation Principles and Priorities

The Belt and Road Initiative abides by the principles of the UN Charter and the five principles of Peaceful **Coexistence**. Besides, it also observes the following four principles. Firstly, the Initiative advocates open cooperation. All countries and international organizations can participate in it, so that the joint efforts can benefit more regions. Secondly, the Initiative upholds harmony and inclusiveness. It promotes cultural tolerance and respects the development paths and models chosen by different countries. It seeks the common ground to develop while tolerating differences, so that all countries can live in peace and prosperity. Thirdly, the Initiative follows market rules and international norms. It gives full play to the role of the market in resource distribution and the role of the government in **macro-control**. Finally, the Initiative expects mutual benefit and win-win results. It will fully consider the interests and concerns of all parties and allow full play to their strengths and potentials.

There are five cooperation priorities for all the concerned countries since they have their own resource advantages and their economies are mutually complementary. Therefore, they try to cooperate in order to achieve five major goals: policy coordination, facilities connectivity, unimpeded trade, financial integration, and people-to-people bonds.

China in Action

Achievements have been made since the Chinese government actively promote the construction of the Belt and Road. The positive **connotation** of the Initiative have been interpreted further and broad consensuses have been reached through the official communication between Chinese government and the concerned countries.

On the one hand, many projects about cooperation framework have been signed. China have signed a memorandum for the construction of Belt and Road with Tajikistan, Kazakhstan, Qatar and Kuwait. In addition, a series of key cooperation projects have been promoted in areas, such as infrastructure connectivity, industrial investment and resource development. During President Xi Jinping's visit to Pakistan, projects with a total of 46 billion dollars have been signed. And during his visit to Indonesia, an agreement on the cooperation and construction of the high-speed railway has been signed. On the other hand, various domestic resources has been coordinated and the policy support has been strengthened. For example, the **Asian Infrastructure Investment Bank**[1] has been established and the Silk Road Fund has been launched. Meanwhile, a series of international **summits**, forums, **seminars** and expos on the Belt and Road have been successfully held across the country, playing an important role in enhancing understanding, reaching consensus and deepening cooperation.

The Belt and Road Initiative is one of the most significant and far-reaching ones that China has ever put forward. Its idea carries forward the spirit of the ancient Silk Road, which was based on equality, mutual trust and benefits. It also conforms to the 21st century norms of promoting peace, development, cooperation and adopting a win-win strategy for all. Therefore, the Belt and Road Initiative is not only a strategic conception for the great **rejuvenation** of the Chinese nation, but also a beneficial path for the common prosperity of countries along the proposed route and a combination of the "Chinese dream" with the global dream of stability and prosperity.

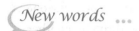

New words

initiative [ɪˈnɪʃətɪv] *n.* An initiative is an important act or statement that is intended to solve a problem. （重要的）法案；倡议

allocation [ˌæləˈkeɪʃn] *n.* The allocation of something is the decision that it should be given to a particular person or used for a particular purpose. 分配；分派

integration [ˌɪntɪˈgreɪʃn] *n.* The integration of something means that two things become closely linked or form part of a whole idea or system. 整合；一体化

optimize [ˈɒptɪmaɪz] *v.* To optimize a plan, system, or machine means to arrange or design it so that it operates as smoothly and efficiently as possible. 使优化

coordinate [kəʊˈɔːdɪneɪt] *v.* If you coordinate an activity, you organize the various people and things involved in it. 协调

excavate [ˈekskəveɪt] *v.* 挖掘

comprehensive [ˌkɒmprɪˈhensɪv] *adj.* Something that is comprehensive includes everything that is needed or relevant. 全面的

sustainable [səˈsteɪnəbl] *adj.* You use sustainable to describe the use of natural resources when this use is kept at a steady level that is not likely to damage the environment. 可持续的

cohesive [kəʊˈhiːsɪv] *adj.* Something that is cohesive consists of parts that fit together well and form a united whole. 统一的

infrastructure [ˈɪnfrəstrʌktʃə] *n.* The infrastructure of a country, society, or organization consists of the basic facilities such as transport, communications, power supplies, and buildings, which enable it to function. （国家、社会、组织赖以行使职能的）基础建设，基础设施

complementary [ˌkɒmplɪˈmentri] *adj.* Complementary things are different from each other but make a good combination. 相互补充的；相辅相成的

corridor [ˈkɒrɪdɔː] *n.* A corridor is a strip of land that connects one country to another or gives it a route to the sea through another country. 走廊（通往他国或经由他国通向大海的狭长地带）

coexistence [ˈkoɪgˈzɪstəns] *n.* The coexistence of one thing with another is the fact that they exist together at the same time or in the same place. 共存

macro-control [ˈmækrəʊ kənˈtrəʊl] *n.* 宏观调控

connotation [ˌkɒnəˈteɪʃn] *n.* The connotations of a particular word or name are the ideas or qualities which it makes you think of. 内涵意义

summit [ˈsʌmɪt] *n.* A summit is a meeting at which the leaders of two or more countries discuss important matters. 峰会

seminar [ˈsemɪnɑː] *n.* A seminar is a meeting where a group of people discuss a problem or topic. 研讨会；专题讨论会

rejuvenation [rɪˌdʒuːvəˈneɪʃn] *n.* 振兴

Phrases and expressions

adhere to　遵循
free-flow　自由流动；自由流
comply with　遵从
abide by　遵守
participate in　参加
give full play to　充分发挥

Chapter 7　Silk and Ceramics Culture

Proper names

Silk Road Economic Belt　丝绸之路经济带
21st Century Maritime Silk Road　21世界海上丝绸之路
New Eurasia Land Bridge　新欧亚大陆桥
Mongolia　蒙古
Indochina Peninsula　中南半岛
Pakistan　巴基斯坦
Bangladesh　孟加拉国
Myanmar　缅甸

Cultural Notes

1. **Asian Infrastructure Investment Bank**（AIIB）is a multilateral development bank that aims to support the building of infrastructure in the Asia-Pacific region, to promote the construction of regional connectivity and strengthen the cooperation of China and other Asian countries and regions. The bank is proposed as an initiative by the government of China and is headquartered in Beijing. Till 2017, it has 77 member states.

Exercises

1. Read the following statements and try to decide whether it is true or false according to your understanding.

____ 1) In September 2013, when President Xi Jinping visited Central Asia and Southeast Asia, he raised the Initiative of jointly building the Silk Road Economic Belt and the 21st-Century Maritime Silk Road.

____ 2) The Initiative is designed to enhance the efficient allocation of resources, further market integration and establish a framework for regional economic cooperation for all concerned countries.

_____ 3) Focusing on the five routes, the Belt and Road will take advantage of international transport routes as well as core cities and key ports to further strengthen collaboration and build five international economic cooperation corridors.

_____ 4) The Belt and Road Initiative aims to connect Asia, Europe and Africa along five routes.

_____ 5) Achievements have been made since the Chinese government actively promote the construction of the Belt and Road. For example, China have signed a memorandum for the construction of Belt and Road with Tajikistan, Kazakhstan, Qatar and Kuwait.

2. Fill in the blanks with the information you learn from the text.

1) The Belt and Road Initiative refers to the _____ and _____.

2) The Initiative is designed to enhance _____, further market integration and establish a framework for regional economic cooperation for all concerned countries.

3) The Belt and Road Initiative abides by the principles of the UN Charter and the five principles of _____.

4) The countries along the Belt and Road try to cooperate in order to achieve the five major goals: policy coordination, facilities connectivity, unimpeded trade, financial integration, and _____.

5) During Xi Jinping's visit to _____, an agreement on the cooperation and construction of the high-speed railway has been signed.

3. Explain the following terms.

1) The Belt and Road Initiative
2) Asian Infrastructure Investment Bank

4. Translate the following paragraph into English.

进入21世纪，在以和平、发展、合作、共赢为主题的新时代，面对复苏乏力的全球经济形势、纷繁复杂的国际和地区局面，传承和弘扬丝绸之路精神更显重要和珍贵。2013年9月和10月，中国国家主席习近平在出访中亚和东南亚国家期间,先后提出共建"丝绸之路经济带"和"21世纪海上丝绸之路"的重大倡议,得到国际社会高度关注。加快"一带一路"建设,有利于促进沿线各国经济繁荣与区域经济合作,加强不同文明交流互鉴,促进世界和平发展,是一项造福世界各国人民的伟大事业。

5. Critical thinking and discussion.

The "Maritime Silk Road" is the sea lane through which the ancient Chinese and other parts of the world will make economic and cultural exchanges. In the middle and late Tang Dynasty, because the land Silk Road was blocked by the war, the maritime Silk Road reached an unprecedented boom. In 2013, when Chinese President Xi Jinping made a speech in Kazakhstan, the construction of Silk Road Economic Belt and the Maritime Silk Road was proposed.

How do you think to spread Chinese civilization while pushing economic cooperation between China and the countries along the Belt and Road?

Chapter 8
Chinese Tea Culture

Section A Origin and Development of Tea

China is the homeland of tea. China had tea shrubs as early as five to six thousand years ago. Tea, along with silk and porcelain, began to be known to the world over more than a thousand years ago and has since always been an important Chinese export. Of the three major beverages of the world-tea, coffee and cocoa, tea is consumed by the largest number of people in the world. People regard it as the bond of friendship and the symbol of civilization. It has become a national drink in China. It is not only one of necessities of Chinese people's daily life, but the best medium of cultural exchange. It is not exaggerating to say that the magic of tea has **condensed** the cultural essence of Chinese nation. The discovery and **utilization** of tea is a great contribution made by Chinese nation to human beings.

Origin of Tea

As for the origin of tea, tea experts have proved that it originated in the southwest of China. In Yunnan Province and elsewhere, there are still some wild tea trees that are over 1,000 years old. Tea in fact started as a kind of medicine in China to treat some illness of human being. Besides, the **folklore** is also available. It is said that tea was firstly discovered and tasted by Emperor Shennong, who was commonly regarded as one of the oldest forefathers and the inventor of Chinese herbal medicine and agriculture.

There is a well-known story of *Shennong Tasting A Hundred Plants*. One day, after walking for a long time, Shennong felt tired and thirsty, so he rested under a tree and started a fire to boil water in a pot. Suddenly some leaves fell into the pot from a nearby tree. Shennong drank the water and found it not only sweet and tasty but **freshening** as well. He felt less tired, so he went on to drink all the water from the pot.

Another version of this tale is a little different and more amazing. It is said that Shennong tried 72 different kinds of poisonous plants in a day and he lay on the ground, barely alive. At this moment, he noticed several rather **fragrant** leaves dropping from the tree beside him. Out of curiosity and habit, Shennong put the leaves into his mouth and chewed them. After a little while he felt well and energetic again. So he picked more leaves to eat and thus **detoxified** his body from the poisons. These legendary leaves are said to be the earliest tea.

Whatever the story, tea interested Shennong and attracted him to do further research into its characteristics. The ancient Chinese medical book called ***Shennong Herbal Medicine***, which is attributed to him, states that "Tea tastes bitter. Drinking it, one can think quicker, sleep less, move more **nimbly**, and see more clearly". This was the earliest book to record the **medicinal** effects of tea.

Development of Tea

In Shennong era, tea was favored for its medical usage. By Zhou Dynasty, the function of tea to refresh body and clear mind had gradually replaced its function as medicine. People started drying the leaves to preserve tea. When they made tea, they put the leaves into a pot and made a kind of thick tea soup. The princes of Zhou Dynasty liked to drink this thick soup, but due to its bitterness, it did not become widely popular.

In the end of the Western Han Dynasty, people began to cultivate tea plants **purposively**. A typical example was the **cultivation** of tea garden in Mengshan Mount, Sichuan Province. The tea cultivated here was so popular that it became **tribute** tea. It was said that only 360 tea shoots would be picked, rolled, dried and then stored in two silver pots as tribute to be sent to Chang'an city (now Xi'an city), where the emperor and the royals would use it as **sacrifice** in ceremonies praying for good harvests. So it was called "the first tea in the world". During this period, tea was highly expensive and usually only available for the emperor and other high ranking nobles. Meanwhile, tea drinking and tea making technology gradually spread from Sichuan to other provinces in the south.

It was said that the flourishing of tea planting could be to a large extent attributed to **Buddhism**, and many traditional famous teas in China were created by monks. During the period of the Wei, Jin, and the Northern and Southern Dynasties, Buddhism was popular in our country. As was known to all, Buddhism **advocated** long time **meditation** and **apprehension** of Buddhist doctrines, and tea happened to be refreshing and good for purifying the heart and **diminishing** the desire. Therefore, the demand for tea in temples promoted tea planting. Besides, during this period, **literati** formed a habit of taking medicinal pills and elixirs（丹药）for **immortality**. Since tea was considered to have the magical effect of health preservation and **longevity**, it was appreciated by many of them, which in turn contributed to the popularity of tea drinking. With the increase of tea planting year by year, tea gradually became popular among ordinary people in the south. "Sit to have a cup of tea" has become a hospitality **etiquette**.

Tang Dynasty was a **milestone** in the development of tea industry in ancient China. In this period a method of tea processing called "green steaming" （蒸青）was invented, with the aim to rid tea leaves of their "grassy" flavor. After steaming, the tea leaves were made into the shape of pies, and then dried and sealed for storage. It was also in Tang Dynasty that teahouses in their true senses came into being, and in some big cities there were also tea shops, which stored a great many tea leaves and prepared tea for their customers. Poems and articles **dedicated** to tea also appeared, and many poets wrote about poems on tea such as **Yuan Zhen**, **Bai Juyi**, **Liu Yuxi**, etc.

Tang Dynasty's **prominence** was not only in the production and marketing of tea, but also in the development of tea culture in this period. Tang Dynasty saw the first **definitive** publication about tea—*Cha Jing*[1], which was the first of its kind in the world. This book contained a comprehensive summary of all aspects of tea culture, including medicinal uses, tea picking, making and cooking, as well as its **utensils**. It was a complete **synthesis** of knowledge about tea at that time. Its author, **Lu Yu**[2], was consequently dubbed the "Saint of Tea" by later generations. *Cha Jing* established the earliest traditional tea science and laid the foundation for the development of Chinese tea culture.

During this period, tea also spread to China's frontier of **Uyghur**, Tibet and other minority areas. People in these regions ate a large amount of mutton or beef in their diets, and tea had the effect of pro-moting **digestion** and reducing **greasiness**, so it became the daily necessities for the minority people. But the ethnic regions, due to climate reasons, couldn't produce tea, and then the "tea-horse market" in history appeared, which

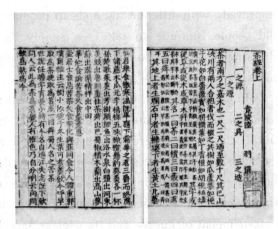

Cha Jing

gradually became the prosperous routine of *chamagudao* (茶马古道), also called the Ancient Tea-Horse Road. What's more, tea became the most popular **commodity** in foreign trades, and Japanese Buddhists brought tea seeds from China to Japan.

Song Dynasty was a golden age for tea. The royal and intellectuals' **affections** for tea further promoted its popularity. The calligrapher **Cai Xiang** wrote *Record of Tea* and Emperor **Zhao Ji** wrote *General Remarks on Tea*. Also, more tea types appeared and tea **connoisseurs** even held competitions to judge the quality of tea in terms of the tea leaves, water and the mixture. Tea became people's essential drink and teahouse played a prominent role in daily life. To obtain a fair profit, the government fully carried out tea monopoly, or *que cha* (榷茶) policy, setting up six tea **monopoly** centers and thirteen tea plantations in tea planting and commercial regions to take control of tea production and trade. The country benefited a lot from tea taxes.

Different from former governments, the Ming government allowed people to plant tea bushes freely. Tea culture, which had been set back by Mongolians, underwent a **renaissance** as dark tea, green tea, and oolong tea all developed during this time. **Zhu Yuanzhang**, the first Ming Emperor, ordered that the tribute tea be changed from compressed tea to loose tea, and this tradition has been **retained** ever since.

As their knowledge of tea broadened, tea farmers were no longer content to harvest tea from the wild, but began to pay more attention to planting and cultivating tea bushes, while at the same time tea processing techniques improved, with different methods to produce the six major types of tea. Nor did people continue to take tea simply as food or medicine. Rather, tea drinking began to take on a spiritual dimension, containing deep cultural meanings.

In Qing Dynasty, tea had truly reached to the folk life and been widely loved by all walks of life. People also paid attention to **exquisite** teapots. Yixing purple clay teapots were the most popular during this period. Teahouses were everywhere on the street. The number of the tea-related works, including books, poems and paintings, was the biggest of all dynasties. For example, there were detailed descriptions of tea evaluation and appreciation, its use for entertainment of guest, and its uses as sacrifice and engagement presents or gifts for friends in novels, like *The Dream of Red Mansion*, *The Travels of Lao Can*, etc. In terms of commerce, the Qing government did business with many countries in Asia, Africa and Europe, such as Britain and Holland.

With the popularity of tea, tea shops and teahouses are springing up in China cities. Tea becomes a national drink in China and plays an **irreplaceable** part in Chinese life. And Chinese tea culture and arts also attract a lot of foreign tourists to China. With the development of cultural exchange, commerce and trade between China and other countries, tea has already traveled all over the world. Now more than one-half of the population of the world consume tea. The daily consumption of tea is approximately 3 billion cups all over the world. Since 1990s, tea production in China has been over 20% of the world's total output, and the export volume of tea has exceeded 15% of the world's total exports. Now China is becoming one of the largest tea producing, exporting and consuming countries in the world.

New words

condense [kən'dens] *v.* If you condense something, especially a piece of writing or speech, you make it shorter, usually by including only the most important parts. 压缩；精简（尤指文章或讲话）

utilization [ˌjuːtəlaɪˈzeɪʃn] *n.* Utilization is the act of using something. 利用；使用

folklore [ˈfəʊklɔː(r)] *n.* Folklore is the traditional stories, customs, and habits of a particular community or nation. 民间传说；民俗

freshening [ˈfreʃnɪŋ] *adj.* If you freshen something up, you make it clean and pleasant in appearance or smell. 使清新；使清爽或鲜亮

fragrant [ˈfreɪɡrənt] *adj.* Something that is fragrant has a pleasant, sweet smell. 气味芬芳的；芳香的

detoxify [ˌdiːˈtɒksɪfaɪ] *v.* If you detoxify, or if something detoxifies your body, you do something to remove poisonous or harmful substances from your body. 清除（体内）的毒素；（使）解毒

nimbly [ˈnɪmblɪ] *adv.* Someone who is nimble is able to move their fingers, hands, or legs quickly and easily. 敏捷的；灵活的；灵巧的

medicinal [məˈdɪsɪnl] *adj.* Medicinal substances or substances with medicinal effects can be used to treat and cure illnesses. 有疗效的；药用的；治病的

purposively [ˈpɜːpəsɪvli] *adv.* 有目的地；故意地

cultivation [ˌkʌltɪˈveɪʃn] *n.* The act of raising or growing plants (especially on a large scale). 种植；栽培

tribute [ˈtrɪbjuːt] *n.* (especially in the past) money given by one country or ruler to another, especially in return for protection or for not being attacked. 贡品；贡金

sacrifice [ˈsækrɪfaɪs] *n.* the act of offering sth to a god, especially an animal that has been killed in a special way. 祭祀；祭品

Buddhism [ˈbʊdɪzəm] *n.* Buddhism is a religion which teaches that the way to end suffering is by overcoming your desires. 佛教

advocate [ˈædvəkeɪt] *v.* If you advocate a particular action or plan, you recommend it publicly. 提倡；主张

meditation [ˌmedɪˈteɪʃn] *n.* Meditation is the act of remaining in a silent and calm state for a period of time, as part of a religious training, or so that you are more able to deal with the problems of everyday life. （宗教中）打坐；冥想

apprehension [ˌæprɪˈhenʃn] *n.* The apprehension of something is awareness and understanding of it. 理解；领悟

diminish [dɪˈmɪnɪʃ] *v.* When something diminishes, or when something diminishes it, it becomes reduced in size, importance, or intensity. （使）减小；（使）减弱；（使）降低

literati [ˌlɪtəˈrɑːti] *n.* Literati are well-educated people who are interested in literature. 文人学士

immortality [ˌɪmɔːˈtæləti] *n.* The quality or state of being immortal. 不死；长生

longevity [lɒnˈdʒevəti] *n.* Longevity is long life. 长寿；长命

etiquette [ˈetɪket] *n.* Etiquette is a set of customs and rules for polite behaviour, especially among a particular class of people or in a particular profession. （尤指特定阶层的）礼仪，礼节；（尤指特定行业的）行为规范，规矩

milestone [ˈmaɪlstəʊn] *n.* A milestone is an important event in the history or development of something or someone. 里程碑；重大事件；重要阶段

dedicate [ˈdedɪkeɪt] *v.* If someone dedicates something such as a book, play, or piece of music to you, they mention your name, for example in the front of a book or when a piece of music is performed, as a way of showing affection or respect for you. 把（书、戏剧、音乐作品等）献（给）

prominence [ˈprɒmɪnəns] *n.* If someone or something is in a position of prominence, they are well-known and important. 重要；杰出；著名

definitive [dɪˈfɪnətɪv] *adj.* A definitive book or performance is thought to be the best of its kind that has ever been done or that will ever be done. （书籍或表演）最佳的；最具权威的

utensil [juːˈtensl] *n.* Utensils are tools or objects that you use in order to help you to cook or to do other tasks in your home. 器皿；用具

synthesis [ˈsɪnθəsɪs] *n.* A synthesis of different ideas or styles is a mixture or combination of these ideas or styles. 综合；结合；结合体

digestion [daɪˈdʒestʃən] *n.* Digestion is the process of digesting food.

消化

greasiness [ˈgriːzɪnɪs] *n.* Consisting of or covered with oil. 多脂；油腻

commodity [kəˈmɒdəti] *n.* A commodity is something that is sold for money. 商品

affection [əˈfekʃn] *n.* If you regard someone or something with affection, you like them and are fond of them. 喜爱；喜欢

connoisseur [ˌkɒnəˈsɜː(r)] *n.* A connoisseur is someone who knows a lot about the arts, food, drink, or some other subject. （艺术、食品、酒等的）鉴赏家；鉴定家；行家

monopoly [məˈnɒpəli] *n.* If a company, person, or state has a monopoly on something such as an industry, they have complete control over it, so that it is impossible for others to become involved in it. 垄断；专营服务

renaissance [rɪˈneɪsns] *n.* If something experiences a renaissance, it becomes popular or successful again after a time when people were not interested in it. 复兴；复活；复燃

retain [rɪˈteɪn] *v.* To retain something means to continue to have that thing. 保留；保持；保存

exquisite [ɪkˈskwɪzɪt] *adj.* Something that is exquisite is extremely beautiful or pleasant, especially in a delicate way. 精美的；精致的

irreplaceable [ˌɪrɪˈpleɪsəbl] *adj.* Irreplaceable things are so special that they cannot be replaced if they are lost or destroyed. （因为特殊而）不能替代的；独一无二的

Phrases and expressions

along with　　连同；以及；和⋯一起
attribute to　　把某事归因于；认为某事［物］属于某人［物］
set back　　推迟；拨慢
spring up　　迅速成长

Proper names

Shennong Herbal Medicine 《神农本草》，现存最早的中药学著作约起源于神农氏，代代口耳相传，于东汉时期集结整理成书，成书非一时，作者亦非一人，是秦汉时期众多医学家搜集、总结、整理当时药物学经验成果的专著，是对中国中医药的第一次系统总结。

Yuan Zhen （唐）元稹，诗人。创作了杂言诗《一字至七字诗·茶》，一字至七字诗俗称宝塔诗，在中国古代诗中较为少见。

Bai Juyi （唐）白居易，诗人。白居易终生、终日与茶相伴，白居易一生写了二千多首诗，提及茶事的有六十三首之多，应居唐朝诗人之冠。

Liu Yuxi （唐）刘禹锡，诗人。在唐代诗人里被称为"诗豪"，对茶的感觉是很敏感的，特别对茶与诗、茶与酒的关系。他说，"诗情茶助爽，药力酒能宣。"有了茶，作诗有了灵感，刘禹锡的诗好与茶有很大关系。长寿也和茶有关系，七十一岁寿终也算是古来稀了。

Uyghur （地名）维吾尔。维吾尔族主要聚居在新疆维吾尔自治区，主要分布于天山以南，塔里木盆地周围的绿洲是维吾尔族的聚居中心，其中尤以喀什噶尔绿洲、田绿洲以及阿克苏河和塔里木河流域最为集中。天山东端的吐鲁番盆地，也是维吾尔族较为集中的区域。

Cai Xiang （宋）蔡襄，书法家、文学家、政治家和茶学家，撰写了《茶录》。蔡襄促进了北苑茶的发展，福建茶叶在北宋时期名列首位，应功归于蔡襄。

Zhao Ji （宋）赵佶，即宋徽宗，对中国茶事的最大贡献是撰写了中国茶书经典之一的《大观茶论》，为历代茶人所引用。其中尤其是关于点茶的一篇，详细记录了宋代流行点茶这种代表性的茶艺，为后人了解宋代点茶提供了依据。

Zhu Yuanzhang （明）朱元璋，即明太祖。明朝初期，贡茶仍然采用福建的团饼，后来，朱元璋认为，进贡团饼茶太"重劳民力"，决意改制，下令罢造"龙团"，改进芽茶。明太祖的诏令，在客观上，对进一步破除团饼茶的传统束缚，促进芽茶和叶茶的蓬勃发展，起到了有力的推动作用。

The Dream of Red Mansion 《红楼梦》，中国古典四大名著之首，清代作家曹雪芹创作的章回体长篇小说，又名《石头记》《金玉缘》。书中有关茶的内容就有493处之多，其中包括茶俗、茶礼、茶诗词、茶叶、茶具、泡茶用水、泡茶方法、品茶环境、茶疗方剂、用茶禁忌等，可谓是我国历代文学作品中记述与描绘得最全、最生动的。

The Travels of Lao Can 《老残游记》，清末文学家刘鹗的代表作。书中最具特色的是对北方景物、风俗民情的描写，如第九回"三人品茗促膝谈心"描绘了平民百姓在山中喝野茶的情景。

Cultural Notes

1. ***Cha Jing*** （《茶经》）or *The Classic of Tea* is the first known monograph on tea in the world, by Chinese writer Lu Yu during the Tang Dynasty. Lu Yu's original work is lost; the earliest editions available date to the Ming Dynasty. The book is not large, about 7,000 Chinese characters in a condensed, refined and poetic style of literary Chinese language. It is made of "Three Scrolls and Ten Chapters" （三卷十章）.

2. ***Lu Yu*** （陆羽，733-804）is respected as the Sage of Tea for his contribution to Chinese tea culture. He is best known for his monumental book *The Classic of Tea*, the first definitive work on cultivating, making and drinking tea.

Exercises

1. Read the following statements and try to decide whether it is true or false according to your understanding.

____ 1) Of the three major beverages of the world—tea, coffee and cocoa, tea is consumed by the largest number of people in the world.

____ 2) It is said that tea was firstly discovered and tasted by Emperor Huangdi.

____ 3) It was said that the flourishing of tea planting could be to a large extent attributed to Taoism.

____ 4) In Song Dynasty, tea monopoly, or *quecha* policy was fully carried out.

____ 5) Zhu Yuanzhang, the first Ming Emperor, oversaw a change from compressed tea to loose tea, and this tradition has been retained ever since.

2. Fill in the blanks with the information you learn from the text.

1) What can be roughly confirmed is that tea originated in the _____ of China.

2) _____, was consequently dubbed the "Saint of Tea" by later generations.

3) In _____, tea had truly reached to the folk life and be widely loved by all walks of life.

4) _____, a method of tea processing, was invented in Tang Dynasty.

5) Tea in fact started as a kind of _____ in China to treat some illness of human being.

3. Explain the following terms.

1) The legend of Shennong and tea

2) The development of tea in Tang Dynasty

4. Translate the following paragraph into English.

"你要茶还是咖啡?"是用餐人常被问到的问题,许多西方人会选咖啡,而中国人则会选茶。相传,中国的一位帝王于5 000年前发现了茶,并用来治病。在明清(the Ming and Qing Dynasties)期间,茶馆遍布全国,饮茶在6世纪传到日本,但直到18世纪才传到欧美。如今,茶是世界上最流行的饮料(beverage)之一。茶是中国的瑰宝,也是中国传统和文化的重要组成部分。

5. Critical thinking and discussion.

It was reported that tea farmers in Ya'an, Sichuan Province used panda droppings to fertilize tea. Such tea was healthy because pandas absorb just 30 percent of the nutrients from their diet of wild bamboo, and pass on the remaining 70 percent to plants via the high-quality organic compost. The special panda tea would be sold for 220,000 yuan ($35,051) for just 500 grams, making it one of the most expensive tea in the world. What is your comment on this report?

Section B Tea Varieties and Benefits

There are fifteen major tea-producing provinces in mainland China, and Taiwan also produces tea. According to different ways of processing, especially the extent of **fermentation**, tea is usually divided into three basic types: green tea, oolong tea and black tea. **Alternatively**, based on the ways of processing and the characteristic qualities of manufactured products, tea is generally classified into six major types: green tea, black tea, oolong tea, white tea, yellow tea and dark tea. Each type has its representative famous tea, each with its unique appearance and aroma, and some are even associated with beautiful legends. More importantly, different types of tea have common and unique benefits to human health. The best teas, well-known for their top quality in color, fragrance and taste, are obtained under the following situations, such as excellent natural growing conditions, top-class tea trees, refined picking methods, and exquisite processing techniques.

Green tea is the oldest type of tea in China, and it is also produced in the largest quantity. Many provinces and cities are renowned for their production of green tea, the most **eminent** provinces being Zhejiang, Jiangxi and Anhui. The leaves of green tea are not fermented, so they largely retain the original flavor of tea, which is simple, elegant and **enduring**. At first sip, green tea may taste a little light, but after a while it gains a fragrance in the mouth that lingers. To make green tea, the methods used are mainly (green) steaming, (green) frying and (green) sun-drying, to remove the moisture in the fresh tea leaves and to bring out their fragrance. There are well-known varieties, such as West Lake Longjing Tea of Zhejiang Province, Biluochun Tea of Jiangsu Province, Huangshan Maojian Tea of Anhui Province, Xinyang Maojian Tea of Henan Province, Duyun Maojian Tea of Guizhou Province, and Laoshan Green Tea of Shandong Province, etc.

Longjing Tea   Biluochun Tea

Black tea accounts for over 90% of all tea sold in the West and it is also the most popular form of tea consumed in South Asia. Black tea is full-fermented tea. In the process of fermentation, not only the leaves turn black, but its soup becomes bright red. Meanwhile, black tea forms fruit aroma and unique mellow taste. While green tea retains the light and refreshing flavor of the leaves, the fermented black tea gives a stronger and thicker flavor. Black tea first appeared in Qing Dynasty, and so it is much younger than green tea. Black tea originated in Fujian and its **vicinities**, and later spread to other provinces in the south. According to different processing methods and characteristics, Chinese black tea is divided into **Gongfu black tea**, **Xiaozhong black tea** and crushed black tea. Among them, *Gongfu* black tea, noted for its fine and delicate production, is a special product of China. In Chinese and the languages of neighboring countries, black tea is known as "red tea", a description of the color of the liquid; the Western term "black tea" refers to the color of the oxidized leaves. The best brands of black tea are Qihong (祁红) from Anhui Province and Dianhong (滇红) from Yunnan Province.

Dianhong Black Tea   Qihong Black Tea

Chapter 8 Chinese Tea Culture

Oolong tea is a **specialty** from southeastern China, originating from provinces of Fujian, Guangdong and Taiwan. Oolong tea features a partial fermentation process, and thus has the characteristics of both green and black teas. It tastes as clear and fragrant as green tea and as strong and refreshing as black tea. Different varieties of oolong tea can vary widely in flavor. They can be sweet and fruity with honey aromas, or woody and thick with roasted aromas, or green and fresh with complex aromas, all depending on the **horticulture** and the style of production. Different varieties of oolong tea are processed differently, but the leaves are usually formed into one of two distinct styles. Some are rolled into long curly leaves, while others are wrap-curled into small beads, each with a tail. The former style is more traditional. Also, high quality oolong teas produce a long aftertaste that lingers in mouth. Being semi-fermented, oolong tea is quite effective in breaking down protein and fat, aiding weight loss. In recent years it has been welcomed by more and more people with the name of "weight losing tea" and "bodybuilding tea". **Tieguanyin**, **Dahongpao** and **Wuyi Rock Tea** from Fujian as well as **Dongding Oolong Tea** from Taiwan are among the most prized oolong teas.

Tieguanyin Tea

DahongPao Tea

Dark tea was invented by accident. Long ago, in order to supply tea to the ethnic groups of the northwest, tea produced in Yunnan, Sichuan, Hubei, Hunan, and other places had to be transported to the north, and then to the northwest via the Silk Road. On horseback, the tea travelled far and was affected by the weather, and **alternating** damp and dry conditions caused major changes to the chemical composition of the leaves, and also turned them **blackish**-brown. In spite of this, they still gave off a rare fragrance, and this type of tea quickly came to be appreciated by the locals. Dark tea is a type of fermented tea, known in China as *hei cha* (黑茶). The fermentation process, which may last from several months to many years, is extremely **exacting**. After the unique process, the finished tea takes on a dark brown color. Unlike most Chinese teas whose taste and aroma fade with age, fermented tea can actually be aged to improve its flavor. The fermented leaves last much longer than other types of tea. As a Chinese specialty, dark tea is usually compressed into different shapes for storage and transport convenience.

In the past, dark tea was the most exported tea in China, which was shipped as far away as Russia. It is also the most popular tea in areas of China with large ethnic minority populations. People from Tibetan, Mongolian and Uyghur ethnic groups consider fermented tea as an essential part of their daily lives. The most famous brand of this variety is the **Pu'er Tea** from southwest China's Yunnan Province.

Pu'er Raw Tea   Pu'er Aged Tea

Famous as a medicinal tea, it is believed to aid digestion, reduce **cholesterol**, lower blood pressure, reinforce the **immune** system and help prevent cancer. The smooth and dark Pu'er tea has a rich and distinctively **earthy** flavor.

Yellow tea is an increasingly rare and expensive variety of tea. The process for making yellow tea is similar to that of green tea but with an added step of **encasing** and steaming the tea. This allows the tea to **oxidize** at a slower rate, producing a far more mellow taste than is found in most green teas; this also gives the leaves a slightly yellow coloring during the drying process. Yellow tea is often placed in the same category with white tea due to its light **oxidation**. One of the primary aims of making yellow tea is to remove the characteristic grassy smell of green tea while preserving the health qualities of green tea. Some have even **speculated** that yellow tea may be healthier because it is easier on the stomach than green tea. Famous yellow tea includes **Junshan Silver Needle Tea** (Hunan Province), **Huoshan Yellow Tea** (Anhui Province), **Pingyang Huangtang Tea** (Zhejing Province), **Dayeqing Tea** (Guangdong Province), etc.

Junshan Silver Needle Tea

Huoshan Yellow Tea

White tea derives its name from the distinctive white-colored appearance of the dry tea. It is made with uncured buds and young leaves of some tea **cultivars** from Fujian Province. These buds and leaves go through minimal processing so that they are kept closer to their natural state. Even the silvery-white hairs (茶毫) on the leaves are preserved, which gives the dry tea a **whitish** appearance. Both green and white teas are among the most lightly oxidized teas, which increases the tea's **antioxidant** properties. Young tea leaves contain higher **caffeine** than old ones, so the caffeine content of white tea may be higher than that of green tea. China's white tea sells well in the United States because American scientists found that elements from white tea are beneficial to people's health. Well-known brands of white tea are **Baihao Silver Needle** and **White Peony**.

Baihao Silver Needle Tea

White Peony Tea

In addition to the above six major types of tea, there are also **tight-pressed teas** and scented teas produced by re-processing. Tight-pressed tea comes from tight pressing crude tea leaves after steaming at a high temperature. This kind of tea can be divided into cake tea, brick tea, roll tea and other groups according to their shape. *Tuo* tea (沱茶) of Yunnan is an outstanding example of this type. Scented tea, with a history of over 1,000 years, it is made from a mixture of **edible** flowers and tea leaves. Scented teas were popular in Qing Dynasty. Jasmine tea, the most common scented tea, is very popular in Beijing and Tianjin. In recent years, there are teabag, instant tea, bottled or canned tea, herbal tea and other **therapeutic** tea with various ingredients added.

In a word, tea not only has good flavors but also benefits to health, so it is loved by many people at home and abroad. Different kinds of tea have different functions which have positive effects on our health. For example, green tea can **dispel** the effects of alcohol, black tea warm stomach, and dark tea reduce blood pressure and help lose weight. Oolong tea does good for body building and dieting. In all, tea has great medicinal values, such as anti-cancer, lowering blood pressure, improving eyesight and **restraining** disease, reducing stress and so on.

New words

fermentation [ˌfɜːmenˈteɪʃn] *n.* A process in which an agent causes an organic substance to break down into simpler substances. 发酵

alternatively [ɔːlˈtɜːnətɪvli] *adv.* You use alternatively to introduce a suggestion or to mention something different to what has just been stated. 要不然；或者

eminent [ˈemɪnənt] *adj.* unusual; excellent. 非凡的；杰出的

enduring [ɪnˈdjʊərɪŋ] *adj.* lasting for a long time 持久的；耐久的

vicinity [vəˈsɪnəti] *n.* If something is in the vicinity of a particular place, it is near it. 邻近地区；附近

specialty [ˈspeʃəlti] *n.* A specialty of a particular place is a special food or product that is always very good there. （某个地方的）特色食品；特产

horticulture [ˈhɔːtɪkʌltʃə(r)] *n.* Horticulture is the practice of growing plants. 园艺

alternate [ɔːlˈtɜːnət] *v.* When you alternate two things, you keep using one then the other. When one thing alternates with another, the first regularly occurs after the other. （使）交替；（使）轮流

blackish [bˈlækɪʃ] *adj.* Something that is blackish is very dark in color. 带黑色的；深色的

exacting [ɪɡˈzæktɪŋ] *adj.* You use exacting to describe something or someone that demands hard work and a great deal of care. 费劲的；苛刻的；要求严格的

cholesterol [kəˈlestərɒl] *n.* Cholesterol is a substance that exists in the fat, tissues, and blood of all animals. Too much cholesterol in a person's blood can cause heart disease. 胆固醇

immune [ɪˈmjuːn] *adj.* If you are immune to a particular disease, you cannot be affected by it. 免疫的

earthy [ˈɜːθi] *adj.* If you describe something as earthy, you mean it looks, smells, or feels like earth. 泥土般的；泥土味的

encase [ɪnˈkeɪs] *v.* If a person or an object is encased in something, they are completely covered or surrounded by it. 把…包住；把…封起来

oxidize [ˈɒksɪdaɪz] *v.* When a substance is oxidized or when it oxidizes, it changes chemically because of the effect of oxygen on it. （使）氧化

speculate [ˈspekjuleɪt] *v.* If you speculate about something, you make guesses about its nature or identity, or about what might happen. 推测；猜测；猜想

oxidation [ˌɒksɪˈdeɪʃən] *n.* Oxidation is a process in which a chemical substance changes because of the addition of oxygen. 氧化

cultivar [ˈkʌltɪvɑː(r)] *n.* A variety of a plant developed from a natural species and maintained under cultivation. 栽培品种

whitish [ˈwaɪtɪʃ] *adj.* Whitish means very pale and almost white in color. 发白的；苍白的

antioxidant [ˌæntiˈɒksɪdənt] *n.* An antioxidant is a substance which slows down the damage that can be caused to other substances by the effects of oxygen. Foods which contain antioxidants are thought to be very good for you. 抗氧（化）剂；阻氧化剂；防老（化）剂

caffeine ['kæfiːn] *n.* Caffeine is a chemical substance found in coffee, tea, and cocoa, which affects your brain and body and makes you more active. 咖啡因

edible ['edəbl] *adj.* If something is edible, it is safe to eat and not poisonous. 可食用的；能吃的

therapeutic [ˌθerə'pjuːtɪk] *adj.* Therapeutic treatment is designed to treat an illness or to improve a person's health, rather than to prevent an illness. 治疗性的；有助于治疗的

dispel [dɪ'spel] *v.* To dispel an idea or feeling that people have means to stop them having it. 驱散；消除

restrain [rɪ'streɪn] *v.* To restrain something that is growing or increasing means to prevent it from getting too large. 抑制；限制；控制

Phrases and expressions

account for （在数量、比例上）占
fade with 褪去，失去光泽；逐渐消逝
derive from 由…起源；取自

Proper names

Gongfu black tea　功夫红茶
Xiaozhong black tea　小种红茶
Tieguanyin　铁观音

Chapter 8 Chinese Tea Culture

> **Dahongpao**　大红袍
> **Wuyi Rock Tea**　武夷岩茶
> **Dongding Oolong Tea**　冻顶乌龙茶
> **Pu'er Tea**　普洱茶
> **Junshan Silver Needle**　君山银针
> **Huoshan Yellow Tea**　霍山黄茶
> **Pingyang Huangtang**　平阳黄汤
> **Dayeqing**　大叶青
> **Baihao Silver Needle**　白毫银针
> **White Peony**　白牡丹
> **tight-pressed tea**　紧压茶

Exercises

1. Read the following statements and try to decide whether it is true or false according to your understanding.

_____ 1) Tieguanyin and Dahongpao are representatives of dark tea.

_____ 2) Green tea is the oldest type of tea in China.

_____ 3) Black tea is partially fermented tea.

_____ 4) The process for making yellow tea is similar to that of green but with an added step of encasing and steaming the tea.

_____ 5) Scented teas were popular in Ming Dynasty.

2. Fill in the blanks with the information you learn from the text.

1) According to the extent of fermentation, tea is usually divided into three basic types: _____, oolong tea and black tea.

2) Biluochun Tea of _____ Province is a type of green tea.

3) _____ accounts for over 90% of all tea sold in the west and it is also the most popular form of tea consumed in South Asia.

4) _____ features a partial fermentation process, and thus has the characteristics of both green and black teas.

5) _____ is a type of fermented tea, known in China as *hei cha*.

3. Explain the following terms.

1) Green tea

2) Oolong tea

4. Translate the following paragraph into English.

功夫茶（*gongfu* tea）不是一种茶叶或茶的名字，而是一种冲泡的手艺。人们叫它功夫茶，是因为这种泡茶方式十分讲究：它的操作过程需要一定的技术，以及泡茶和品茶的知识和技能。功夫茶起源于宋朝，在广东的潮州府（今潮汕地区）一带最为盛行，后来在全国各地流行。功夫茶以浓度（concentration）高著称。制作功夫茶主要使用的茶叶是乌龙茶（oolong tea），因为它能满足功夫茶色、香、味的要求。

5. Critical thinking and discussion.

How to cultivate love for tea among young people in China?

Section C Tea Customs and Etiquette

Among all 56 nationalities in China, no matter whether they belong to farming culture or **nomadic** culture, almost all of them have the habit of drinking tea. Although different ethnic groups have different tea customs, it is common for hosts to serve a cup of tea to guests to show hospitality and respect, while not providing tea is often regarded impolite.

Tea Customs

It is very popular for Chinese people to treat guests with tea. Traditionally, a visitor to a Chinese home is expected to sit down and drink hot tea while talking. For the guest, not to take at least a sip might be considered rude or offensive in some areas. Chinese people can chat with a friend for a whole afternoon over a pot of good tea. In modern China, even the simplest **dwelling** has a tea set and a water heater for making a hot cup of tea. These **implements** are symbols of welcome visitors and neighbors.

Also it is quite common for many Chinese families to drink tea after dinner, so to speak tea is part of their living and working. This not only helps promote digestion but also provides a good time for family gathering and communication.

Besides, Chinese younger generation offer a cup of tea to show their respect to the elder generation. This is especially popular during big celebrations, such as birthdays or the Spring Festival. Today, sometimes parents may pour a cup of tea for their children to show their care, or a boss may even pour tea for **subordinates** to promote their relationship. However, the lower ranking person should not expect the higher ranking to serve him or her the tea on formal occasions.

In addition, tea plays a significant part in both engagements and weddings for Chinese people. Tea is an important element of engagement gifts given to the bride's family by the groom's family before the wedding. At the wedding, tea ceremony is one of the most significant events. It's called *jing cha* （敬茶）in Chinese, literally meaning, "to respectfully offer tea". It

Jing Cha at the Wedding

incorporates a very formal introduction of the bride and the groom, and expresses respect to their families. In the distant past, the young couple were required to kneel while serving tea. Nowadays, most families only require the couple to bow. This is a way of expressing gratitude to their elders for all the years of love and care. On some occasions, the bride serves the groom's family, and the groom serves the bride's family. This process symbolizes the joining together of the two families.

Tea may also be offered as part of a formal apology. For example, children who have misbehaved may serve tea to their parents as a sign of regret and **submission**.

In modern China, tea is an important social tool. A teahouse is the **by-product** of Chinese tea culture. Teahouse has been ranked as the public place for drinking tea, relaxation and entertainment since ancient times, acting as a vivid **epitome** of Chinese tea culture and Chinese people's leisure lives. Chinese people generally consider meeting in a teahouse to be a good opportunity to socialize or discuss business matters. **Conventionally**, Chinese scholars prefer to have free and deep communication with their friends, and businessmen usually conduct negotiations with their business partners, while enjoying some tea. People go to teahouses, not only for the drink, but also for a place to meet people.

Teahouse

It cannot be denied that over a long time people in different regions and of different nationalities developed their own unique customs of drinking tea. In Guangdong, for example, people like drinking morning tea, in Fujian they prefer *gongfu* tea, Hunan has *lei* tea, Sichuan people love **covered-bowl tea**, while people of the Bai nationality treat their guests with **three-course tea**. Tibetan people prefer buttered tea and those from Inner Mongolia like milk tea. These various tea customs constitute the rich and profound Chinese tea culture.

Tea Ceremony and Etiquette

Tea ceremony or tea art, called *chayi* (茶艺) in Chinese, is the professional show of Chinese tea culture and the way for brewing and drinking tea. It also has a long history. The habit of drinking tea in China started during Zhou Dynasty. The skill of brewing and serving tea was regarded important as early as Han Dynasty. Today, the tea ceremony is a popular cultural activity not only welcomed by Chinese but also by foreigners.

Tea ceremony consists of the skills for judging and commenting tea, the artistic procedures for brewing tea and the atmosphere for enjoying tea. A whole tea ceremony generally includes the following procedures: tea selection, water selection, tea-brewing technique and environmental **refinement**. The background of tea ceremony is also very important, because it is usually the best way to show the deep thought of tea culture.

During tea ceremony, a real drinker is strict about the usage of water, tea type and utensils as well as the atmosphere selection. Drinking tea traditionally can raise personal morality and help to harmonize interpersonal relations. Tea drinkers usually like making friends and showing their personalities by virtue of drinking tea. In a word, understandably, drinking tea is the presentation of elegance, personality, status, educational background, religious belief and cultural pursuit.

Tea drinking today is usually streamlined into a simpler ceremony. It may be carried out in several ways, for example, *gaiwan shi*（盖碗式）, covering the cup style, and *gongfu shi*（功夫式）, skilful style. *Gaiwan shi* is the simplest because only a tea cup with its cover is used to contain the tea and the tea drinker simply sips the tea and enjoys it. *Gongfu shi* is the most **authentic** as it has its origin from Lu Yu's **treatise**.

Dating back to Tang Dynasty, *gongfu* tea ceremony is now the most famous type of Chinese tea ceremony and popular in **Chaoshan area** (Chaozhou City, Shantou City and Jieyang City) in Guangdong Province. The most important element of *gongfu* tea is the tea sets. There are at least ten tea sets for Chaoshan *gongfu* tea ceremony. The way to practice *gongfu* tea ceremony includes five basic steps:

Gongfu Tea Sets

Step 1, prepare a bottle of boiled water.

Step 2, put the tea leaves into the tea cup with hot water and soak for about 30 seconds, and then spill the water.

Step 3, put the tea leaves into the tea **funnel** to filter out the **impurities**.

Step 4, pour the hot water again, and use the cup lid to stir the tea leaves a little bit. Step 5, pour the tea into the tea funnel again, and it is ready to drink.

Generally, a teacup would be filled to seven-tenths of its capacity. It is said that the other three-tenths would be filled with friendship and affection. In the drinking process, after the drinker's cup is filled again that person may knock his or her bent index and middle fingers (or some similar variety of finger tapping) on the table to express gratitude to the person who serves the tea.

There is also some serving etiquette in each way of tea ceremony. Take *gaiwan shi* for example. First, hosts are supposed to serve tea with two hands holding the saucer and bow slightly forward. Make sure the guests don't have to move or stand up to receive the *gaiwan*. Second, those receiving the tea should hold the saucer but not the cup as the *gaiwan* cup itself can be hot. Third, they should hold the saucer to move the cup close to their mouths, which is the most traditional way of drinking tea from *gaiwan*. Finally, once they've drunk the tea, take back the *gaiwan* with two hands, once again by holding the saucer.

Gaiwan Tea sets

In terms of tea sets, there are no fixed regulations, but people do follow some common practices. Green tea goes with transparent glass or white porcelain; scented tea with *gaiwan*, transparent glass, or white porcelain with a cover; *gongfu* black tea goes with purple clay ware with white inside glaze, or with white porcelain or warm colored wares or coffee wares. And oolong tea is also excellent in purple clay ware. In a word, the harmonious combination of function, material, and color of tea ware is essential to brewing excellent tea.

Unlike the world-renowned Japanese tea ceremony, the Chinese one emphasizes the tea rather than the ceremony, like taste of the tea and the difference between the teas in various cups. Moreover, Chinese tea ceremony stresses the harmonious, peaceful, optimistic and authentic atmosphere. Tea culture is a kind of **intermediate** culture which passes on the spirit of traditional Chinese culture to the future generation. Together with China's influence and cultural communication overseas, Chinese tea and tea culture spread widely in the world.

Chapter 8　Chinese Tea Culture

New words

nomadic [nəʊˈmædɪk] *adj*. Nomadic people travel from place to place rather than living in one place all the time. 游牧的；游牧部落的

dwelling [ˈdwelɪŋ] *n*. A dwelling or a dwelling place is a place where someone lives. 住宅；住所；居所

implement [ˈɪmplɪmənt] *n*. An implement is a tool or other piece of equipment. 工具；器具；用具

subordinate [səˈbɔːdɪnət] *n*. If someone is your subordinate, they have a less important position than you in the organization that you both work for. 下级；下属；部属

incorporate [ɪnˈkɔːpəreɪt] *v*. If one thing incorporates another thing, it includes the other thing. 包含

submission [səbˈmɪʃn] *n*. Submission is a state in which people can no longer do what they want to do because they have been brought under the control of someone else. 屈服；投降；归顺

by-product [baɪ ˈprɒdʌkt] *n*. Something that is a by-product of an event or situation happens as a result of it, although it is usually not expected or planned. 附带产生的结果

epitome [ɪˈpɪtəmi] *n*. If you say that a person or thing is the epitome of something, you are emphasizing that they are the best possible example of a particular type of person or thing. 典型；缩影

conventionally [kənˌvenʃənəli] *ad*. in a conventional manner 通常的；传统的

refinement [rɪˈfaɪnmənt] *n*. Refinements are small changes or additions that you make to something in order to improve it. Refinement is the process of making refinements. 完善；修正；改进

authentic [ɔːˈθentɪk] *a*. An authentic person, object, or emotion is genuine. 真的；真正的；真诚的

treatise [ˈtriːtɪs] *n*. A treatise is a long, formal piece of writing about a particular subject. （专题）论文

funnel [ˈfʌnl] *n*. A funnel is an object with a wide, circular top and a narrow short tube at the bottom. Funnels are used to pour liquids into containers which have a small opening, for example bottles. 漏斗

impurity [ɪmˈpjʊərəti] *n*. Impurities are substances that are present in small quantities in another substance and make it dirty or of an unacceptable quality. 杂质；不纯物质

intermediate [ˌɪntəˈmiːdiət] *a*. An intermediate stage, level, or position is one that occurs between two other stages, levels, or positions. 居中的；中间的

Phrases and expressions

rank as　把…看作；可算作…
by virtue of　凭借…的力量
pass on　传递

Proper names

lei tea　擂茶。擂者，研磨也。擂茶，就是把茶叶、芝麻、花生等原料放进擂钵里研磨后冲开水喝的养生茶饮。擂茶在中国华南六省都有分布。保留擂茶古朴习俗的地方有：湖南的桃源、临澧、安化、桃江、益阳、凤凰、常德等地，广东省的揭西、清远、英德、陆河、惠来、五华等地；江西省的赣县、石城、兴国、于都、瑞金等地；福建省的将乐、泰宁、宁化等地；广西的贺州黄姚、公会、八步等地；台湾的新竹、苗栗等地。

covered-bowl tea　盖碗茶。在汉族、回族居住的大部分地区都有喝盖碗茶的习俗，盖碗茶盛于清代，如今，在四川成都、云南昆明等地，已成为当地茶楼、茶馆等饮茶场所的一种传统饮茶方法，一般家庭待客，也常用此法饮茶。盖碗茶是一种上有盖、下有托、中有碗的茶具，又称"三才碗"，盖为天、托为地、碗为人。

three-course tea 三道茶。三道茶也称三般茶，是云南白族招待贵宾时的一种饮茶方式。驰名中外的白族三道茶，以其独特的"头苦、二甜、三回味"的茶道早在明代时就已成了白家待客交友的一种礼仪。

Chaoshan area 潮汕地区，简称潮、潮汕，广东省辖地，是汕头、潮州和揭阳三市的统称，也称潮州地区，史称潮州八邑。潮汕地处中国东南沿海，闽、粤、台三省交界，中国大陆海岸线与北回归线相交处。

Exercises

1. Read the following statements and try to decide whether it is true or false according to your understanding.

_____ 1) Chinese tea ceremony emphasizes the tea rather than the ceremony, like the taste of the tea and the difference between the teas in various cups.

_____ 2) Tea is an important element of engagement gifts given to the bride's family by the groom's family before the wedding.

_____ 3) People of the Miao nationality treat their guests with three-course tea.

_____ 4) *Gaiwan* tea ceremony is the most famous type of Chinese tea ceremony and popular in Chaoshan area.

_____ 5) Generally, a teacup would be filled to seven-tenths of its capacity.

2. Fill in the blanks with the information you learn from the text.

1) For the guest, not to take at least a sip might be considered _____ in some areas.

2) _____ should not expect the higher ranking people to serve him or her tea on formal occasions.

3) *Jingcha* at the wedding ceremony symbolizes _____ .

4) _____ has been ranked as the public place for drinking tea, relaxation and entertainment since ancient times, acting as a vivid epitome of Chinese tea culture and Chinese people's leisure lives.

5) _____ , called *chayi* （茶艺） in Chinese, is the professional show of Chinese tea culture and the way for brewing and drinking tea.

3. Explain the following terms.

1) *Gong fu* tea ceremony
2) Etiquettes in *gaiwan* tea ceremony

4. Translate the following paragraph into English.

中国是一个文化历史悠久的（time-honored）国家，也是一个礼仪（ceremony and decorum）之邦。当客人来访的时候，一般都要泡茶给客人喝。在给客人奉茶之前，你应该询问一下他们都喜欢喝什么类型的茶，并采用最合适的茶具奉上。在奉茶期间，主人需要仔细留意客人茶杯里的茶量。通常，若是用茶杯泡的茶，在茶喝完一半之后就应该加开水，这样，茶杯就能保持一直都是满的，茶的芳香（bouquet）也可以保留。

5. Critical thinking and discussion.

Can you recommend some ways to spread Chinese tea culture throughout the world?

Chapter 9
Chinese Alcohol Culture

Section A Origin and Development of Chinese Alcohol

Drinks in the world fall into two types, non-alcoholic and alcoholic. The former includes plain water, milk, juice, coffee, tea and soft drink and the latter is primarily classified into beer, wine, cider and **spirits**. In China, alcohol, also called *jiu*（酒）, has won the favor of people from all walks of life, and developed its own unique culture such as legends about its origin over thousands of years. Among them, some beautiful stories have been widely **circulating** up to now. With the collective efforts of ancient Chinese, the brewing techniques have dramatically improved. For example, the types of distiller's yeasts（酒曲）have been greatly **diversified** over time. The category of alcohol has got enriched and famous liquor（名酒）has come out such as guizhou *maotai*（贵州茅台）. It has even developed its drinking **etiquette** and its inspiration to literati. As a shining pearl in Chinese civilization, Chinese alcohol has been well-accepted in the world.

Legends about Origin

Some hold that the origin of alcohol can be traced back to Shennong era because at that time the **cultivation** of millet（小米）in the middle reaches of the Yellow River provided materials for the production of alcohol. While others believe that it appeared in the Yellow Emperor period. As it is recorded in the medical classic ***Plain Questions*** that the Yellow Emperor and his chancellor Qi Bo（岐伯）discussed how to brew alcohol with glutinous millet, millet, wheat, beans and rice.

Among multiple versions about the origin of alcohol, two have received wide acceptance. They are closely related to two Chinese gods of wine: Yi Di (仪狄) and Du Kang (杜康). One goes that Yi Di, a female winemaker, was asked to make a mellow alcohol to **relieve** the **fatigue** of Yu the Great (大禹). She tried many times and finally succeeded in making some **fermented** glutinous rice wine. After one sip, Yu felt refreshed and **intoxicated**. However, he banned it shortly after sobering up for he feared that some successors would **overindulge** themselves in it. And the other gives credit to Du Kang who was generally believed to be a minister of the Yellow Emperor or a ruler of Xia Dynasty, Shao

Du Kang

Kang. He was assumed to produce *shujiu* (秫酒), an alcohol made mainly from sorghum (高粱), and was honored as the **deified** patron of winemakers in China. Du Kang was so highly recognized in China that his name became a **byword** for any good alcohol, as is shown in "What is it that can disperse worries? It's Du Kang alcohol", lines out of *A Short-Song Ballad* by Cao Cao, a warlord in the late Eastern Han Dynasty. These two versions indicate that alcohol could relieve fatigue and drown sorrows.

Development of Alcohol

Discoveries of late Stone Age jugs suggest that fermented drinks existed at least as early as the New Stone Age. And Patrick McGovern, an anthropologist of the University of Pennsylvania Museum, said that the earliest chemically **confirmed** alcoholic liquid in the world was discovered at Jiahu (贾湖), Henan Province, back to 7000-6600 BC. Then he added that the highly fermented liquid came naturally out of wild grapes, hawthorns and honey in proper conditions. In ancient times, trees **thrived** and were laden with fruits. Some of them fell down, clustered and fermented gradually into the natural alcohol if the temperature and humidity permitted. The natural phenomenon inspired the ancient people to explore the man-made brewing. They loaded up wild fruits, animal milk and other sugary substances into containers and closely observed the process of fermentation. After numerous trials, they gained a good understanding of the necessary conditions of fermentation such as temperature and humidity. Then they applied this knowledge to the fermentation of grains. Hence the variety of alcohol has increased.

In ancient times, people started to put *qunie*（曲蘗）, the earliest starter culture（发酵剂）, into the grains to speed up the fermentation. In Shang Dynasty, winemakers separated *qu* (**mildewed** grains) and *nie* (**malted** grains). The liquid made from *qu* was called *jiu* with 15%-20% alcohol content, while that made from *nie* was named *li*（醴）with about 4% alcohol content. Then in Northern Song Dynasty, they trialed the red starter（红曲）. It created a pigment coloring the liquor in **shades** of red. Different distiller's yeasts

Li

and wine-making methods brought about the increase in alcohol variety. And many famous liquors came out especially since Tang Dynasty, such as *jiannan shaochun*（剑南烧春）, *bingtangchun*（冰堂春）and *xifengjiu*（西凤酒）. Among them, *jiannan shaochun* was **designated** as a **tribute** to the imperial court, sharing the same origin with today's *mianzhu daqu*（绵竹大曲）and *jiannanchun*（剑南春）.

In long-term wine-making practice, ancient Chinese discovered that alcohol differed with water in boiling point, then they could get liquid with higher alcohol **concentration** by **distillation**. This technique together with **sophisticated** fermentation has helped Chinese alcoholic drinks gain international reputation for being excellent, exquisite and **aromatic**. For instance, *guizhou maotai* won worldwide fame when being awarded a gold medal at the 1915 **Panama-Pacific Exposition** in San Francisco, and has been served at state banquets with foreign heads of state and distinguished guests visiting China. In America, some young people favor Chinese *white spirit* and vividly call it "firewater".

Classification

Baijiu (*white spirit*), *huangjiu* (*yellow wine*), wine and beer are the four series of alcohol in China. As one of the six world-famous varieties of spirits (the other five being brandy, whisky, rum, vodka, and gin), *baijiu* is a distilled alcoholic drink with a relatively high alcohol content. It's made from various grains such as sorghum, millet, corn, rice and wheat. And it's renowned for being crystal clear, aromatic and tasty. In terms of its fragrance, *baijiu* is primarily classified into 6 categories: sauce fragrance, strong fragrance, light fragrance, rice fragrance, phoenix fragrance (凤香型) and mixed fragrance (兼香型). It is usually stored at least several years to develop its full fragrance and flavor before hitting the market. As one of the best-known brands of *baijiu*, *guizhou maotai* usually has at least four years of storage. Thus it produces a pure, mild, and mellow sauce fragrance, and is highly recognized at home and abroad. Besides, other famous liquors such as *wuliangye* (五粮液), *fenjiu* (汾酒), *gujinggongjiu* (古井贡酒), *jiannanchun* (剑南春), *luzhoulaojiao* (泸州老窖) and *xifengjiu* (西凤酒) are also noted for their full fragrance and flavor after long-term storage.

Guizhou *Maotai*

As one of the three ancient alcoholic drinks in the world (the other two being beer and wine), *huangjiu*, or yellow wine, typically contains less than 20% alcohol by volume. It is also made from various grains such as rice, millet, wheat, sorghum and corn. And it's known for its **transparent** amber color and luster, **balmy** fragrance, and sweet and mild taste. Judging by sugar content, *huangjiu* primarily includes 5 categories: dry,

Jimo Old Wine

semi-dry, semi-sweet, sweet and extra-sweet. It is usually warmed up to 60-70℃ before drinking. Such temperature can help give off its aroma. Jimo Rice Wine, Fujian Rice Wine and Shaoxing Rice Wine are well-known *huangjiu*.

Wine, ranging from 5.5% to 15.5% in alcohol content, is mainly made from grapes. It can be divided into red, white and rose red types by color, **sparkling** and **still** by form, or dry, semi-dry, semi-sweet and sweet by sugar content. Its sweetness is determined by the amount of residual sugar relative to the acidity in the wine after fermentation. Although wine was long **overshadowed** by *baijiu* and *huangjiu*, its consumption has grown dramatically since the implement of the reform and opening-up policy. In 1980, French **Rémy Martin** set up a joint venture in China and first produced **Dynasty Wine**. Shortly afterwards, many Chinese brands mushroomed such as **Great Wall**, **Suntime** and **Changyu**. Nowadays they have gradually **dominated** China's domestic wine market and have initiated an **upsurge** in wine drinking. In China, Yantai-Penglai region, the home of Changyu, is the largest wine producing area. It has over 140 wineries and the wine production there accounts for about 40% of the domestic market.

Beer has a long brewing history in China. Archaeological evidences have proved that *li*, light and sweet in flavor, is considered to be the earliest beer, for it bears much resemblance to today's beer in taste and alcohol content. Beer is mainly made from barley and hops (啤酒

Tsingtao Beer

花). Customarily, it is classified into pale, brown and stout beers (黑啤) by color. And according to the world-recognized classification standard, it is divided into Top Fermentation and Bottom Fermentation by yeast. The former refers to the fermentation on the surface of liquid at higher temperature; the latter at the bottom of liquid at lower temperature. Currently, the majority of Chinese beer falls into Bottom Fermentation type. One of its typical example is Tsingtao Beer. It is brewed with spring water from Laoshan Mountain in Qingdao and takes up about 15% of domestic market share. Apart from Tsingtao Beer, other major Chinese brands are Yanjing Beer and Snow Beer of

Beijing, and Zhujiang Beer of Guangzhou.

China's 9,000-year history makes it hard to **verify** the exact beginning of alcohol. This uncertainty adds more charm and mystery to Chinese alcohol culture and also arouses foreigners' fascination with China. Actually, it is inadvisable to owe its invention to any individual, for it is definitely impossible to accomplish such a great feat without collective efforts. Their **superb** craftsmanship has magnificently contributed to the prosperity of Chinese alcohols. The rich variety in raw materials and updated brewing techniques have resulted in the diversification of alcohol category and the improvement of alcohol quality. Without doubt, alcohol has become an essential part of Chinese people's daily life and has a special position in traditional Chinese culture.

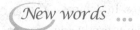

New words

spirits [ˈspɪrɪts] *n.* Spirits are strong alcoholic drinks such as whisky and gin. (威士忌、杜松子酒等) 烈性酒

circulate [ˈsɜːkjəleɪt] *v.* If a piece of writing circulates or is circulated, copies of it are passed round among a group of people. (文章等) 传递；传阅；散发

diversify [daɪˈvɜːsɪfaɪ] *v.* When an organization or person diversifies into other things, or diversifies their range of something, they increase the variety of things that they do or make. 增加…的品种；从事多种经营；(使) 多样化

etiquette [ˈetɪket] *n.* Etiquette is a set of customs and rules for polite behavior, especially among a particular class of people or in a particular profession. (尤指特定阶层的) 礼仪；(尤指特定行业的) 行为规范

cultivation [ˌkʌltɪˈveɪʃn] *n.* If you cultivate land or crops, you prepare land and grow crops on it. 耕；耕作；种植

relieve [rɪˈliːv] *v.* If something relieves an unpleasant feeling or situation, it makes it less unpleasant or causes it to disappear completely. 缓和；

缓解；减轻；解除（不愉快的情绪或形势）

fatigue [fəˈtiːg] *n.* Fatigue is a feeling of extreme physical or mental tiredness. 疲惫；疲劳；疲倦

ferment [fəˈment] *v.* If a food, drink, or other natural substance ferments, or if it is fermented, a chemical change takes place in it so that alcohol is produced. （使）发酵

intoxicated [ɪnˈtɒksɪkeɪtɪd] *adj.* If you are intoxicated by or with something such as a feeling or an event, you are so excited by it that you find it hard to think clearly and sensibly. 陶醉的；极其兴奋的；忘乎所以的

overindulge [ˌəʊvərɪnˈdʌldʒ] *v.* If you overindulge, or overindulge in something that you like very much, usually food or drink, you allow yourself to have more of it than is good for you. 过于沉溺（于）；（通常指）过多地吃（或喝）

deify [ˈdeɪɪfaɪ] *v.* If someone is deified, they are considered to be a god or are regarded with very great respect. 把…神化；把…奉若神明

byword [ˈbaɪwɜːd] *n.* Someone or something that is a byword for a particular quality is well known for having that quality. （某一特征的）代名词；代表人物；典型事物

confirm [kənˈfɜːm] *v.* If something confirms what you believe, suspect, or fear, it shows that it is definitely true. 证实；证明（情况属实）

thrive [θraɪv] *v.* If someone or something thrives, they do well and are successful, healthy, or strong. 兴旺

mildewed [ˈmɪldjuːd] *adj.* Something that is mildewed has mildew growing on it. 生霉的；发霉的

malted [ˈmɔːltɪd] *adj.* Malted barley has been soaked in water and then dried in a hot oven. It is used in the production of whisky, beer, and other alcoholic drinks. （大麦）被制成麦芽的

shade [ʃeɪd] *n.* A shade of a particular color is one of its different forms. For example, emerald green and olive green are shades of green. （色彩的）浓淡；深浅；色度

中国农业文化概览
An Overview of Chinese Farming Culture

designate [ˈdezɪɡneɪt] *v.* If something is designated for a particular purpose, it is set aside for that purpose. 指定

tribute [ˈtrɪbjuːt] *n.* A tribute is something that you say, do, or make to show your admiration and respect for someone. 致敬；颂词；献礼

concentration [ˌkɒnsnˈtreɪʃn] *n.* The concentration of a substance is the proportion of essential ingredients or substances in it. 浓度

distillation [ˌdɪstɪˈleɪʃn] *n.* If a liquid such as whisky or water is distilled, it is heated until it changes into steam or vapor and then cooled until it becomes liquid again. This is usually done in order to make it pure. 蒸馏；用蒸馏法提取

sophisticated [səˈfɪstɪkeɪtɪd] *adj.* A sophisticated machine, device, or method is more advanced or complex than others. （机器、装置等）高级的；精密的；（方法）复杂的

aromatic [ˌærəˈmætɪk] *adj.* An aromatic plant or food has a strong, pleasant smell of herbs or spices. （植物或食品）有香味的；芳香的

transparent [trænsˈpærənt] *adj.* If an object or substance is transparent, you can see through it. 透明的；清澈的

balmy [ˈbɑːmi] *adj.* Balmy weather is fairly warm and pleasant. （天气）温暖舒适的；温和宜人的

sparkling [ˈspɑːklɪŋ] *adj.* Sparkling drinks are slightly fizzy. （饮料）起泡的

still [stɪl] *adj.* Drinks that are still do not contain any bubbles of carbon dioxide. （饮料）不含碳酸气的；不冒泡的

overshadow [ˌəʊvəˈʃædəʊ] *v.* If you are overshadowed by a person or thing, you are less successful, important, or impressive than they are. 使黯然失色；使相形见绌；使显得无足轻重

dominate [ˈdɒmɪneɪt] *v.* To dominate a situation means to be the most powerful or important person or thing in it. 在…中占首要地位；在…中最具影响力

upsurge [ˈʌpsɜːdʒ] *n.* If there is an upsurge in something, there is a sudden, large increase in it. 飙升；急剧增长

verify [ˈverɪfaɪ] *v.* If you verify something, you check that it is true

Chapter 9　Chinese Alcohol Culture

　　by careful examination or investigation. 核实；查证；查清
superb [suːˈpɜːb] *adj*. If you say that someone has superb confidence, control, or skill, you mean that they have very great confidence, control, or skill. （信心、控制力、技巧）非同一般的；超凡的

Phrases and expressions

be classified into　分（类）为
from all walks of life　各行各业的人；各行各业；来自各行各业
sober up　醒酒；清醒起来
drown sorrows　借酒消愁
be laden with　载满
speed up　（使）加速
be renowned for　因…而著名；以…著称
hit the market　上市
give off　发出；散发出
relative to　和…比较起来

Proper names

Plain Questions　《素问》，9卷，81篇，与《灵枢经》为姊妹篇，合称《黄帝内经》。
A Short-Song Ballad　《短歌行》，由东汉末年政治家、文学家曹操所作，其中前八行为"对酒当歌，人生几何！譬如朝露，去日苦多。慨当以慷，忧思难忘。何以解忧？唯有杜康。"
Panama-Pacific Exposition　巴拿马太平洋博览会
Rémy Martin　法国人头马
Dynasty Wine　王朝葡萄酒，由法国人头马有限公司与中国于1980年在

天津建立的合资企业生产。

Great Wall 长城葡萄酒，由中国长城葡萄酒有限公司生产。最早使用"长城"牌葡萄酒的是民权葡萄酒厂，但该商标由中粮酒业有限公司于 1988 年注册。

Suntime 新天葡萄酒，由成立于 1998 年的新天国际葡萄酒业股份有限公司生产。

Changyu 张裕葡萄酒，由爱国人士张弼士于 1892 年创办的张裕酿酒公司生产。

Exercises

1. Read the following statements and try to decide whether it is true or false according to your understanding.

____ 1) The development of alcohol has experienced from distillation to fermented drinks.

____ 2) In Shennong era, the cultivation of millet in the middle reaches of the Yellow River made it possible to produce alcohol.

____ 3) In Northern Song Dynasty, the great progress in alcohol was the development and application of the Red Starter.

____ 4) Yellow wine is famous for its yellow color and luster and the alcohol content is usually 20%-25%.

____ 5) Du Kang, acknowledged to be the founder of wine-making industry, produced *Shujiu*, a real juicy alcohol.

2. Fill in the blanks with the information you learn from the text.

1) The cultivation of _____ in the middle reaches of the Yellow River paved the way for the production of alcohol.

2) Ancient Chinese loaded wild fruits, animal milk and other _____ into containers and closely observed the process of fermentation.

3) Ancient Chinese discovered that alcohol and water had different boiling points, then they began to increase the _____ by means of distillation.

4) Chinese *baijiu* is world-famous for being _____, aromatic and _____.

5) The sweetness of yellow wine is determined by the amount of residual _____ relative to the _____ in the wine after fermentation.

3. Explain the following terms.

1) *Baijiu*

2) *Huangjiu*

4. Translate the following paragraph into English.

茅台酒（*maotai* liquor）是与苏格兰威士忌、法国科涅克（Cognac）白兰地齐名的世界三大蒸馏名酒之一，它独产于贵州省茅台镇。茅台镇的气候、土壤和水质成就了茅台独特的口感。清代时期，茅台成为第一个大规模生产、年产量达170吨的中国白酒。1915年茅台在巴拿马万国博览会（the Panama-Pacific Exposition）上荣获金质奖章，第一次赢得了国际声誉。中华人民共和国成立两年后，茅台被指定为国酒，自此茅台一直用于外国元首和贵宾来访中国时官方场合的宴会。

5. Critical thinking and discussion.

Ma Yun, the founder and executive chairman of Alibaba Group, said he had been longing for *maotai* and more respected its producers than ever before since his personal visit. It must be a miracle to have made the best liquor in Maotai town, Guizhou Province, a remote and inaccessible place in China.

How do you comment this phenomenon that some good wine（名酒）such as *maotai* are made not in some developed areas but in some economically underdeveloped areas? What is your opinion? And why?

Section B Drinking Customs

China is claimed to be a nation of ceremonies. Many rules of etiquette have been well demonstrated in drinking since the Western Zhou Dynasty, and are still **observed** today. For example, the host is supposed to fill the glasses for his guests first before he fills his own. Drinking etiquette is a way to show Chinese **hospitality.** The host always tries hard to prepare a big meal and urges his guests to drink as much as possible, for fear that he might **cold-shoulder** them, even a little bit. Furthermore, in order to create a more cheerful drinking atmosphere, Chinese usually play drinkers' wager game (酒令) in different forms.

Drinking Etiquette

In China, people are inclined to enhance friendship, expand social connections and make negotiations at table. After Guests are seated, according to status and **seniority** and dishes are served, the host offers the first 3 or 6 toasts with some opening remarks, which varies on different occasions. Then the person,

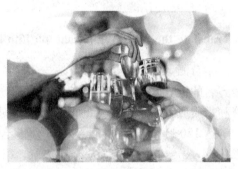

Toast

usually sitting opposite to the host, proposes the same or fewer toasts. After that, the other drinkers are free to offer toasts. They exchange toasts to show their good wishes, sincere respect and heartfelt **gratitude**.

Once a drinker starts proposing a toast, he is supposed to toast to everyone who might **outrank** him at the table, and put himself in an **inferior** status. And it's more courteous if he stands up and empties his glass, while those he drinks a toast to may remain seated or leave their seats to dry up their glasses or sip a little. However, it seems to be a great disgrace and a lack of politeness to decline a toast without a good reason such as pregnancy or sickness. If he wants to impress the person he toasts, he is highly expected to hold the glass in his right hand, place his left one at the glass bottom, and **tip** it slightly towards the person he is drinking with.

What's more, people at the table tend to **clink** their glasses when offering toasts. On some occasions, the drinkers simply tap the surface of the table with the bottom of their glasses. While on others, they usually stand up and lightly touch each other's glass. When one drinker offers a toast to others, he is strongly suggested to place the brink of his glass lower than that of others.

Chinese people usually show their warmth to their guests by urging them to drink. The more the guests drink, the happier the host would be. Otherwise, the host will feel sorry. In China, many interesting sayings have been passed down generation after generation to urge people to drink more, such as "deep feeling, bottom up" "shallow feeling, take a sip" "do not leave until drunk". Usually these sayings are polite, but sometimes forceful or even punishing, especially when one is late for meals or blurts out improper words. Therefore, it is unlikely that the

Clinking Glasses

drinker declines toasts without a ready and **eloquent** tongue or a reasonable **refusal**. When he really cannot drink or has drunk too much, he is more likely to request someone else to drink for him as a way of **evading** drinking with grace.

Drinkers' Wager Game

Drinking is a means of communication for Chinese whether at birthday party, wedding ceremony, family reunion or other togetherness. Usually, in order to break the ice, people often play a drinkers' wager game, called

Drinkers' Wager Game in *A Dream of Red Mansions*

jiuling. When playing the game, one person is in charge and the others should follow his instructions. If they break the rules or lose the game, they will be punished. *Jiuling*-playing was depicted in many Chinese literary classics, such as **A Dream of Red Mansions** and **Journey to the West**. It brings fun to the drinkers and furthers Chinese alcohol culture.

As an integral part of alcohol culture, *jiuling* has a long history dating back to the Western Zhou Dynasty. It was, at that time, introduced to regulate people's drinking behavior in case of excessive drinking. But during the Warring States Period, *jiuling* **evolved** into a particular way to create more **joyous** atmosphere while drinking. And since Tang Dynasty, it has become popular and **generated** diverse artistic forms.

In terms of drinkers' social status and **literacy**, *jiuling* in ancient times was mainly divided into two types, literary and common. The former was knowledgeably demanding and catered to the taste of the well-educated. It covered a wide range of activities such as connecting idioms, composing **couplets** or verses. While the latter was simple and much enjoyed by the less-educated. It also covered a series of activities such as singing, telling stories or jokes.

Jiuling was so appealing to both refined and popular tastes that some of its activities are still popular today such as dice throwing and drum-and-pass game. The former is a tabletop game. When thrown or rolled from the hand or from a container, the dice will come to rest showing on its upper surface a random integer (整数) from one to six. Those who make an incorrect guess will pay a **forfeit** by taking a cup of drink, singing a song and the like. The drum-and-pass game is another popular drinking game. One beats a drum, and the others will quickly pass a flower or something like that from one to another. When the drummer stops, whoever has the flower or something will drink and even give an on-the-spot performance.

The finger guessing game is another popular one. It involves two players at a time. Both drinkers hold out their fingers **simultaneously** while shouting

Finger Guessing Game

out some words containing an integer from 1 to 10 respectively. If one of them gives the total number of fingers held up by both sides, he wins. The loser takes a drink or does a performance as a punishment. If both respond with the correct answer, then goes another round of game. Amazingly, the words mainly **convey** good wishes, as is shown in the table as follows.

Chinese Version	English Version	Number Indicated
一条龙	Being outstanding	1
哥俩好	Befriending you	2
三结义	Being sworn brothers	3
四季财	Making a fortune in four seasons	4
五魁首	Topping the winners	5
六六顺	Going smoothly	6
七个巧	Being skillful	7
八大仙	Being versatile	8
九连环	Being wise	9
满堂红	Being successful	10

Since ancient times, Chinese have been very particular about drinking etiquette. Whether it was at a public banquet or a family dinner, they have strictly followed those rules, such as modesty, **moderate** drinking and respect for seniority at table. The etiquette has guided the **propriety** of drinkers' speech and behavior sub-consciously, which embodies Confucius' key concept of rites. It has become an **epitome** of the well-ordered society, reflecting the desire of Chinese people for peace and harmony.

observe [əbˈzɜːv] v. If you observe something such as a law or custom, you obey it or follow it. 遵守（法律、习俗等）

hospitality [ˌhɒspɪˈtæləti] n. Hospitality is friendly, welcoming behavior towards guests or people you have just met. 殷勤好客；热情友好

cold-shoulder [ˈkəʊldˈʃəʊldə] v. If one person cold-shoulders another, they give them the cold shoulder. 冷落；慢待

seniority [ˌsiːnɪˈɒrəti] n. A person's seniority in an organization is the importance and power that they have compared with others, or the fact that they have worked there for a long time. 资历；年资

gratitude [ˈɡrætɪtjuːd] n. Gratitude is the state of feeling grateful. 感恩

outrank [ˌaʊtˈræŋk] v. If one person outranks another person, he or she has a higher position or grade within an organization than the other person. 级别高于⋯；职位在⋯之上

inferior [ɪnˈfɪərɪə(r)] adj. If one person is regarded as inferior to another, they are regarded as less important because they have less status or ability. （地位、能力等）低等的；低于⋯的；下级的

tip [tɪp] v. If you tip an object or part of your body or if it tips, it moves into a sloping position with one end or side higher than the other. （使）倾斜；（使）斜侧

clink [klɪŋk] v. If objects made of glass, pottery, or metal clink or if you clink them, they touch each other and make a short, light sound. （使）发出叮当声；（使）叮当作响

eloquent [ˈeləkwənt] adj. A person who is eloquent is good at speaking and able to persuade people. 雄辩的；口才流利的；能言善辩的

refusal [rɪˈfjuːzl] n. Someone's refusal to do something is the fact of them showing or saying that they will not do it, allow it, or accept it. 拒绝；回绝；不接受

evade [ɪˈveɪd] v. If you evade something, you find a way of not doing something that you really ought to do. 规避；逃避

evolve [ɪˈvɒlv] v. If something evolves or you evolve it, it gradually develops over a period of time into something different and usually more advanced. （使）逐步发展；（使）演化

joyous [ˈdʒɔɪəs] adj. Joyous means extremely happy. 快乐的；欣喜若狂的

generate [ˈdʒenəreɪt] *v.* To generate something means to cause it to begin and develop. 造成；引起；导致

literacy [ˈlɪtərəsi] *n.* Literacy is the ability to read and write. 读写能力；识字；有文化

couplet [ˈkʌplət] *n.* A couplet is two lines of poetry which come next to each other, especially two lines that rhyme with each other and are the same length. （尤为押韵等长的）对句；对联

forfeit [ˈfɔːfɪt] *v.* If you forfeit something, you lose it or are forced to give it up because you have broken a rule or done something wrong. （因违反规定等）被没收

simultaneously [ˌsɪməlˈteɪnɪəsli] *adv.* Things which are simultaneous happen or exist at the same time. 同时发生的；同时出现的；同步的

convey [kənˈveɪ] *v.* To convey information or feelings means to cause them to be known or understood by someone. 传达；表达；传递

moderate [ˈmɒdərət] *adj.* You use moderate to describe something that is neither large nor small in amount or degree. 中等的；普通的；适度的；适量的

propriety [prəˈpraɪəti] *n.* Propriety is the quality of being socially or morally acceptable. 适当；正当；妥当；得体

epitome [ɪˈpɪtəmi] *n.* If you say that a person or thing is the epitome of something, you are emphasizing that they are the best possible example of a particular type of person or thing. 典型；缩影

Phrases and expressions

be inclined to 倾向于；偏重；有意
pass down 把……一代传一代；使流传
blurt out 突然说出；脱口而出
break the ice 打破僵局；打破冷场
cater to 迎合

appeal to 对…有吸引力
come to rest 停止移动
be particular about 讲究

Proper names ...

A Dream of Red Mansions 《红楼梦》
Journey to the West 《西游记》

Exercises ...

1. Read the following statements and try to decide whether it is true or false according to your understanding.

_____ 1) When drinking at table, both hosts and guests are expected to observe certain rules of etiquette.

_____ 2) If you propose a toast to others, you should put yourself in a superior position to them.

_____ 3) With a ready tongue or a reasonable refusal, the drinker is far from getting drunk.

_____ 4) Originally *jiuling* was introduced to regulate people's drinking behavior and so ensure that they should observe the drinking etiquette in case of drinking to excess.

_____ 5) In the finger guessing game, the one who shouts out the total number of the fingers stretched out by both sides will win.

2. Fill in the blanks with the information you learn from the text.

1) When thrown from the hand or from a container, the dice will come to rest showing on its upper surface a random integer（整数）from _____ to _____.

2) After guests and hosts are seated according to _____ and _____, the host will offer the first 3 or 6 toasts with some opening remarks.

3) If you propose a toast to a senior person, try to make sure that the brink of your glass is _____ than his to show your _____.

4) Literary *Jiuling* was complicated and catered to the taste of the _____.

5) When the drinker really cannot drink, he is more likely to request someone else to drink for him as a way of evading drinking with _____.

3. Explain the following terms.

1) *Jiuling*
2) Finger Guessing Game

4. Translate the following paragraph into English.

酒令是中国人在饮酒时打破僵局的一种餐桌礼仪。在西周，它被用来规范人们的饮酒行为，防止饮酒过度。但在战国时期，酒令演变成饮酒时助兴的一种特殊方式。自唐代以来，酒令开始流行起来，并形成多种艺术形式。根据饮者的社会地位和文化素养，古代酒令主要分为两种：雅令与通令。总的来讲，酒令雅俗共赏。

5. Critical thinking and discussion.

Jiuling was originally introduced to regulate people's drinking behavior in case of excessive drinking. But in modern life, We often see people playing this game in public places such as restaurants or food stalls (大排档). It's so noisy that they may even affect others' dining mood.

Have you ever met or heard of such an experience? What do you think of this situation? Do you think we should cancel *jiuling*? Or in what way can people play it politely?

Section C Alcohol and Social Life

Alcohol, a popular drink to all ranks of people, has been closely associated with many aspects of social life for thousands of years. It was first used as a must in sacrificial ceremonies. It was deified as sacred liquid and only presented to the heaven, the earth and ancestors. The poem At the Foot of Han (《旱麓》) pointed out that the mellow wine in the jade cup was prepared "to offer, to sacrifice" (以享以祀). Then it was enjoyed by the royals and nobles for the state monopoly on its production and distribution, and even became a means to **consolidate** or struggle for power. With the advancement of brewing techniques, alcohol gradually became **affordable** for ordinary people. Nowadays alcohol has become a common drink and is consumed at any time and place. It is safe to assume that alcohol plays an important role in political, military, literary fields and daily life in China.

Alcohol in Politics

Due to the high state monopoly on alcohol, the **privilege** of drinking was only granted to the ruling class such as emperors and vassals (诸侯). They supplied alcohol to entertain their high-ranking officials on various occasions. In Chinese history, many **household** stories are related to power struggle in the guise of drinking. Among them, a very popular one is about two warlords Xiang Yu and Liu Bang in the late Qin Dynasty. Xiang Yu arranged a feast at Hong Gate, a place in Shaanxi Province today, attempting to kill his rival Liu Bang. After a few rounds of drinks, Xiang Zhuang, one of Xiang Yu's generals, was asked to perform a sword dance with the real purpose to assassinate Liu Bang. But Liu Bang saw through the whole scheme and eventually narrowly escaped. This is the historically well-known **Feast at Hong Gate**[1]. When it comes to consolidating the reign with alcohol, no one

Feast at Hong Gate

could **match** Zhao Kuangyin, the first emperor of the Northern Song Dynasty. One day, he **summoned** his major generals to have a drink in his imperial palace. After several rounds of toast, Zhao deprived them of their military power by requiring them to return home with honor, which greatly strengthened his reign.

Alcohol still plays a vital role in today's politics. In China, some famous alcohols are often used to entertain foreign leaders at state banquets in order to better the diplomatic relations with other countries.

Alcohol in Military Affairs

In ancient times, alcohol drinking was generally forbidden in armed forces, for excessive drinking might reduce their fighting capacity and even **incur** fatal disasters. For example, Zhang Fei, a famous general of Shu Kingdom, was killed by his two subordinates after he got drunk. Sometimes alcohol was used to trap the enemy. In the late Eastern Han Dynasty, General Zhou Yu (周瑜) pretended to be drunk, intentionally giving Cao Cao's counselor Jiang Gan (蒋干) a chance to get the false **intelligence**. It wrote Cao's two major commanders of Water Forces would present Cao's head as a tribute to Zhou Yu. **Deceived** by the intelligence, Cao Cao beheaded them, which partially led to Cao's defeat in the **Battle of Red Cliffs**[2].

Oath of the Peach Garden

However, drinking was permitted on the days of **expedition** and victory because a toast could boost **morale**. In Chinese history, many drinking-related stories in wartime have widely spread. In the Spring and Autumn Period, Gou Jian (勾践), king of Yue State, planned to lead his army to attack the state of Wu. Before his riding off to war, his people presented good wine to him. To improve morale, Gou Jian ordered to pour it into a river and then drank with his soldiers upstream. As a result, the morale was greatly boosted and his army became invincible. This is the well-known

Gou Jian Pouring Rice Wine to Reward Soldiers. In the Western Han Dynasty, General Huo Qubing (霍去病) was **granted** some jars of alcohol by Emperor Wu after he **recovered** the lost land **Hexi**. He ordered to empty them into a well and drank with his soldiers. Tang Dynasty poet Wang Han's ***Liangzhou Lines*** describes a magnificent drinking scene of frontier soldiers for their victory. It goes "They are about to drink the finest wine from Evening Radiance cups when the sudden sounding of the *pipa* urges them forth. Don't scorn those drunken fall upon the battlefield. In ancient days or now, how many return who go to war?"

Alcohol in Literature

Since ancient times, alcohol was closely associated with Chinese literature. ***The Book of Songs***, China's first poetry anthology (诗集), collected 30 alcohol-related poems. Some of them mention drinking containers and drinking etiquette; some account the joyful atmosphere of drinking; while some sing high of the magical after-effect of drinking. It's well-recognized among the scholars that alcohol can inspire their literary creativity. Poet Li Bai of Tang Dynasty is a good case in point. With a jar of wine, he produced numerous excellent poems. This was depicted in his contemporary poet Du Fu's famous poem *Ode to the Eight Immortals* (《饮中八仙歌》). Scholar Han Yu of Tang Dynasty mentioned that literati had a ready pen to create lots of good lines after drinking in his second of four poems *Reflections on Spring* (《感春》). Consequently, literati gained a name in their love for drinking in Chinese history. For example, poet Tao Yuanming of Jin Dynasty always indulged himself in alcohol and even got drunk, as was shown in the lines "All my life I never give up wine, because no drinking is a **dismal** sign" from his poem *Stopping Wine* (《止酒》). Poet Li Qingzhao of Song Dynasty drank so much as that she could not return home, as was described in her Song Ci *Like a Dream* (《如梦令》). Additionally, scholars preferred to convey their ideas and feelings about life. There still spread Poet Du Fu's famous lines "The portals of the rich reek of flesh and wine while frozen bodies lie by the roadside" (朱门酒肉臭，路有冻死骨). These words expressed his discontent over inequality and his **indignation** at the bitter social reality. In *Tune*："Phoenix

Hairpin"(《钗头凤》), Poet Lu You of Song Dynasty composed "pink hands so fine, gold-branded wine" to recall the sweetness with his ex-wife Tang Wan(唐婉). Alcohol was also used to strengthen friendship, which was shown by the popular story **Oath of the Peach Garden**[3] in *Romance of the Three Kingdoms*. As a great treasure in Chinese culture, the literature on alcohol is worth cherishing.

With the maturity of wine-making technique, alcohol has become a daily drink from luxury and drinking has been closely related to people's daily life. Nearly all important occasions are celebrated with alcohol. The Spring Festival, an occasion for family reunions, is usually celebrated through drinking with family members and friends. No wedding ceremony is complete unless the newly-weds drink the **nuptial** cup to their future happiness, which symbolizes their love and **commitment** to each other. Similarly, birthday celebration and the ceremony to mark a newborn's 100th day will be **solemnized** with a toast of wine or spirits. It has become the way to express emotions and ideals, to enhance friendship and communication and to **adjust** human relations. Its unique culture is a window to understand the Chinese nation.

New words

consolidate [kənˈsɒlɪdeɪt] *v.* If you consolidate something that you have, for example power or success, you strengthen it so that it becomes more effective or secure. 加强；巩固

affordable [əˈfɔːdəbl] *adj.* If something is affordable, most people have enough money to buy it. 价格合理的；多数人买得起的

privilege [ˈprɪvəlɪdʒ] *n.* If you talk about privilege, you are talking about the power and advantage that only a small group of people have, usually because of their wealth or their high social class. 特权；优惠

household [ˈhaʊshəʊld] *adj.* Someone or something that is a household name or word is very well known. 家喻户晓的；众人皆知的

match [mætʃ] *v.* If you match something, you are as good as it or equal to it, for example in speed, size, or quality. 比得上；配得上；敌得过

summon ['sʌmən] *v.* If you summon someone, you order them to come to you. 召唤；传唤；召见

incur [ɪn'kɜː] *v.* if you incur sth unpleasant, you are in a situation in which you have to deal with it 招致；遭受；引起

intelligence [ɪn'telɪdʒəns] *n.* Intelligence is information that is gathered by the government or the army about their country's enemies and their activities. 情报；谍报

deceive [dɪ'siːv] *v.* If you deceive someone, you make them believe something that is not true, usually in order to get some advantage for yourself. 欺骗；诓骗

expedition [ˌekspə'dɪʃn] *n.* An expedition is an organized journey that is made for a particular purpose such as exploration. 远征；探险

morale [mə'rɑːl] *n.* Morale is the amount of confidence and cheerfulness that a group of people have. 士气

grant [ɡrɑːnt] *v.* If someone in authority grants you something, or if something is granted to you, you are allowed to have it. 授予；给予

recover [rɪ'kʌvə] *v.* If you recover something that has been lost or stolen, you find it or get it back. 重新拿回

dismal ['dɪzməl] *adj.* Something that is dismal is bad in a sad or depressing way. 忧郁的；凄凉的；令人沮丧的

indignation [ˌɪndɪɡ'neɪʃn] *n.* Indignation is the feeling of shock and anger which you have when you think that something is unjust or unfair. 愤怒；愤慨；义愤

nuptial ['nʌpʃl] *adj.* Nuptial is used to refer to things relating to a wedding or to marriage. 结婚的；婚姻的

commitment [kə'mɪtmənt] *n.* A commitment is something which regularly takes up some of your time because of an agreement you have made or because of responsibilities that you have. 承诺；责任；义务

solemnize ['sɒləmnaɪz] *v.* to perform or hold (ceremonies, etc) in due manner. 隆重庆祝

adjust [ə'dʒʌst] *v.* If you adjust something, you change it so that it is more effective or appropriate. 调整；调节

Chapter 9 Chinese Alcohol Culture

Phrases and expressions

It is safe to assume that　可以肯定地说；可以毫不夸张地说
in the guise of　假借；以…为幌子
see through　看透；看穿；识破
narrowly escape　侥幸逃脱
deprive of　剥夺某人的…
a good case in point　一个很好的例子
reek of　发臭气；散发（难闻的气味）

Proper names

Hexi　河西地区，指今甘肃的武威、张掖、酒泉、敦煌等地，因位于黄河以西，自古称为河西，又因其为夹在祁连山与合黎山之间的狭长地带，亦称河西走廊。
Liangzhou Lines　《凉州词》，唐朝诗人王翰所作。
The Book of Songs　《诗经》，中国第一部诗歌总集，收集了自西周初年至春秋中叶五百多年的诗歌305篇。
Romance of the Three Kingdoms　《三国演义》，元末明初小说家罗贯中所著。

Cultural Notes

1. **Feast at Hong Gate**（鸿门宴），also known as Banquet at Hong Gate, Hongmen Banquet, or Hongmen Feast, was a historical event that took place in 206 BC at Hong Gate outside Xianyang, the capital of Qin Dynasty. The two parties involved were Liu Bang and Xiang Yu, two prominent rebellious leaders. Xiang Yu arranged a banquet and invited Liu

Bang to have a drink. While drinking, Xiang Yu's subordinate Xiang Zhuang pretended to perform a sword dance to entertain the guests but actually sought opportunities to assassinate Liu Bang. Liu Bang saw it through and eventually narrowly escaped. The event was one of the highlights of the Chu-Han Contention. The Feast at Hong Gate is often memorialized in Chinese history, fiction and popular culture.

2. Battle of Red Cliffs (赤壁之战), also known as Battle of Chibi, was a decisive battle in the late Eastern Han Dynasty. It was fought in the winter of 208-209 between the allied forces of warlords Liu Bei and Sun Quan and the numerically superior forces of warlord Cao Cao. The allied victory at Red Cliffs ensured the survival of Liu Bei and Sun Quan, and provided a line of defense that was the basis for the later creation of the two southern states of Shu Han and Eastern Wu.

3. Oath of the Peach Garden (桃园三结义), a fictional event in the historical novel *Romance of the Three Kingdoms* by Luo Guanzhong. In order to restore the reign of Liu family, the three heroes Liu Bei, Guan Yu and Zhang Fei took an oath of fraternity in a ceremony in the Peach Garden. They mixed drops of their own blood from their cut wrists into three bowls of alcohol and drank them. From then on, they became sworn brothers who would later play important roles in the establishment of the state of Shu Han during the Three Kingdoms period. It is also often alluded to as a symbol of fraternal loyalty.

Exercises

1. Read the following statements and try to decide whether it is true or false according to your understanding.

____ 1) Alcohol was first used as a must in sacrifice. In the ceremony, it was presented to the heaven, the earth and ancestors.

____ 2) drinking was permitted on the days of expedition and victory because a toast could boost morale.

Chapter 9 Chinese Alcohol Culture

_____ 3) Alcohol was only one kind of drink, having nothing to do with politics.

_____ 4) In ancient China, alcohol drinking was generally forbidden in armed forces, for excessive drinking might reduce their morale.

_____ 5) Poet Du Fu's famous lines "The portals of the rich reek of flesh and wine while frozen bodies lie by the roadside" expressed his discontent over inequality and hatred at the corrupted societies.

2. Fill in the blanks with the information you learn from the text.

1) Due to the high _____ on alcohol in the early dynastic history, the privilege of drinking was only granted to the ruling class.

2) When it comes to consolidating the reign with alcohol, no one could match _____, the first emperor of the Northern Song Dynasty.

3) To improve _____, Gou jian ordered to pour the good wine into a nearly river and then drank with his soldiers.

4) Alcohol can inspire the _____ of literati after drinking. For example, poet Li Bai of Tang Dynasty was the typical representative.

5) No wedding ceremony is complete unless the newly-weds drink the nuptial cup to their future happiness, which symbolizes their _____ and _____ to each other.

3. Explain the following terms.

1) Feast at Hong Gate
2) Gou Jian Pouring Rice Wine to Reward Soldiers

4. Translate the following paragraph into English.

在中国古代，起初酒被视为神圣的液体，用来祭天祭地祭祖宗，与人们的饮食没有关系。随后，由于国家对酒的生产、流通的垄断，只有王室贵族才能享用。随着酿酒技术和酒的产量的提高，酒已成为一种常见的饮料，与人们的日常生活息息相关。自古以来，文人爱酒，酒激发他们的文学创造力。但在动乱时期，一些文人借酒浇愁或沉湎于酒来避祸。

5. Critical thinking and discussion.

Cultural scholar Yu Qiuyu（余秋雨）, in his speech about Chinese spirits culture in 2016, said that when it came to Chinese alcohol, Cao Cao, Tao

Yuanming, Li Bai, Du Fu, Wang Wei and Su Dongpo must have been mentioned. What they said when drinking and the way they drank were impressive.

How do you comment the relationship between alcohol and literati in ancient China? Do you think Chinese young people are as deeply inspired by alcohol as ancient Chinese? And why?

Chapter 10
Chinese Food Culture

Section A Food and Regions

"There is no love sincerer than the love of food", said George Bernard Shaw, an Ireland playwright. It may be more so in China. China is famous for the countless fantastic and delicious dishes all over the world. Its diet is deeply rooted in traditional Chinese culture and has become a unique part in the long history of culture. The **cuisines** in different regions vividly reflect the unique food culture in China and make it full of local color.

China is a vast country with a wealth of local **specialties** for cooking, and in different places, there are different ways for preparing dishes. The formation of a cuisine is closely related to the geographical location, climatic features, local resources as well as the cooking history and eating habit of a specific area. Diets of the north and the south emerged during the Spring and Autumn Period and the Warring States Period. In Tang Dynasty, Chinese cooking got fully developed and the two major culinary styles formed: the south style and the north style. Since then, some local food gradually formed its unique flavor. In the early Qing Dynasty, cuisines of Lu, Su, Yue and Chuan, standing for Shandong, Jiangsu, Guangdong and Sichuan provinces respectively, became the most influential local cuisines, later named the "Four Cuisines". In the late Qing Dynasty, the local specialties of Zhejiang, Fujian, Hunan and Anhui provinces came into being, called Zhe, Min, Xiang and Hui cuisine respectively. Together with the four cuisines, they are known as the "Eight Regional Cuisines". In addition, there are some other cuisines in China, such as the north-eastern China cuisine, Beijing cuisine and Shanghai cuisine, etc.

Lu Cuisine

Lu cuisine has a long history and enjoys great reputation at home and abroad. It mainly consists of Jinan cuisine, cuisine in Jiaodong area (the cities of Qingdao, Yantai and Weihai), and the Confucian-style cuisine. Jinan cuisine is most famous for its soup dishes. Jiaodong dish is **renowned** for cooking seafood with fresh and light taste while the Confucian-style cuisine is well-known for its well-chosen **ingredients** and elegant shaping of the dishes.

Lu cuisine is characterized by its emphasis on aroma, freshness, crispness and tenderness. Scallion and garlic are usually used as seasonings. Hot pepper is also used in some dishes, but spice is less heavily applied than in Sichuan cuisine. The use of soup stock in cooking is another feature of Lu cuisine. Thin soup dishes taste clear and fresh while creamy soup dishes taste thick and strong. Milk soup is also very popular.

The most famous traditional dish in Lu cuisine is Sweet and Sour Carp (糖醋鲤鱼). It is said that the dish originated in the city of Jinan and later gradually spread in the whole country. Since the fish symbol is frequently used to symbolize the wish for more in the sense of good luck, good fortune and long life, Sweet and Sour Carp is commonly served in wedding and birthday feasts as a main **course** to convey

Sweet and Sour Carp

the good wishes. Another well-known dish is Braised Pork Balls in Gravy Sauce, known as *si xi wan zi* (四喜丸子) which means happiness, fortune, longevity and joy (福禄寿喜) in Chinese culture. Other **delicacies** include Braised Intestines in Brown Sauce, Quickly Fried Pig Kidney (爆炒腰花), Braised Sea Cucumber with Scallion (葱烧海参) and so on.

Chuan Cuisine

Chuan cuisine can be traced back to the ancient Ba Kingdom, the modern

Chongqing city, and Shu Kingdom, the modern Chengdu city. It is **distinguished** for its oily, spicy and hot flavor. Spicy and hot foods are the favorite of Sichuan people, because they can help to reduce the moisture inside body caused by the foggy, cloudy and damp climate there.

The taste of Chuan cuisine is quite rich. No wonder people often portray it as "one-hundred dishes, one-hundred flavors". Among them, the most famous are fish-fragrant, sour and spicy, pepper-spicy, spicy-hot, red-oiled and so on. The unique hot taste is created by a mixture of spices and flavorings, including hot pepper, **fermented** bean sauce and Sichuan pepper or *huajiao*（花椒）, resulting in an instantaneous numbing effect on the tongue. Chuan cuisine features quick-frying, stir-frying, dry-frying and dry-braising.

Hot Pot

Hot Pot, one of the most delicious dishes of Sichuan Province, is noted for its spicy, hot, fresh and fragrant taste. It creates a culture of getting together and enjoying good time. In Sichuan, from officials to common people, almost no one does not prefer Hot Pot and all families can make this course. Another notable dish is Sliced Beef and Ox Tongue in Chili Sauce, *fu qi fei pian*（夫妻肺片）in Chinese, which is created by a Chengdu couple. It was selected as the "Appetizer of the Year" of American catering ranking list of the year 2017, which was published by an American magazine in May, 2017. And its name was translated into "Mr And Mrs Smith". Other classic dishes include Mapo Tofu, kung-pao Chicken（宫保鸡丁）, Fish-flavored Pork Slices（鱼香肉丝）, Taibai Duck, to name but a few. Among them, Taibai Duck is not a hot dish in Sichuan cuisine but it's also very famous, because it is closely related to **Libai**[1], a renowned poet in Tang Dynasty.

Yue Cuisine

Yue cuisine is the most widely served style of Chinese cuisine in the world. This is because most of the Chinese who **emigrated** and set up restaurants overseas were from Guangdong province.

Guangdong dishes are characterized by their tender and slightly sweet taste. Sauce is a crucial seasoning in Yue cuisine. The most widely used sauces include hoisin sauce, oyster sauce, plum sauce, and sweet and sour sauce. Other popular **flavorings** include sugar, salt, soy sauce, rice wine and so on. The ingredients of Yue cuisine are very plentiful. Things that are rarely seen on Western tables are commonly used in Yue cuisine such as snake, cat and pangolin (穿山甲). Yue cuisine features braising, stewing and frying to keep the original flavor of the ingredients. Another typical characteristic of Yue cuisine is that there are many kinds of porridge and Dim Sum (点心). Now Dim Sum has been greatly improved by **integrating** the overseas and domestic merits in making it. Every tea house provides a variety of Dim Sum in the city of Guangzhou.

Sweet and Sour Pork with Pineapple

There are many delicacies in Guangdong Province such as Sweet and Sour Pork with Pineapple (菠萝咕咾肉), Steamed Pork with Preserved Greens (梅菜扣肉), Roasted Suckling Pig (烤乳猪) and Salt-roasted Chicken (盐焗鸡).

Min Cuisine

Min cuisine originates from Fujian Province in the south of China. It consists of three styles. Fuzhou style generally tastes light, often with a mixed sweet and sour taste. Western Fujian style features slightly spicy flavoring from mustard and pepper. And Southern Fujian style tastes spicy and sweet.

Min cuisine is known for the use of ingredients from mountain and sea since Fujian province is rich in natural resources. It is also characterized by making various soups. As a saying goes, "it is unacceptable for a meal without soup". Precisely applying various sauces and **seasonings** is another characteristic of Min cuisine. Shrimp oil, soy sauce, brown sugar, crystal sugar, white vinegar, pepper and mustard are widely used to create salty, sweet, sour and spicy tastes. The most characteristic cooking technique is cooking with red rice wine （红曲酒）. The drunken dishes cooked in wine are **prevalent** in Fujian Province as well as other regions throughout China.

Steamed Abalone with Shark's Fin and Fish Maw in Broth, called **Fo tiao qiang**[2] in Chinese, wins renown all over China. Other representative dishes include Sweet and Sour Pork in Shape of Litchi （荔枝肉）, Drunken Chicken （醉糟鸡） and Fuzhou Fish Ball （福州鱼丸） etc.

Steamed Abalone with Shark's Fin and Fish Maw in Broth

Su Cuisine

Su cuisine, popular in the middle and lower reaches of the Yangtze River, is composed of several local styles, most notably Nanjing style, Suzhou style and Yangzhou style. Nanjing style is **reputable** for its fine cutting techniques, which makes the dishes **delicate.** Suzhou style tends to be sweet in taste and excels in using fresh vegetables and seafood as the main ingredients. Yangzhou style, also called Huaiyang style, is well-known for its fine cutting techniques, perfect timing, fine appearance and good quality.

Su cuisine enjoys a reputation for a variety of local ingredients. Its cooking methods are not limited to stir-frying. Stewing, braising, simmering and warming are widely used to preserve the original flavor of ingredients.

In Su cuisine, Boiled Salted Duck （盐水鸭） is a very noted dish. Pure Stewed Meatballs is another well-known dish, which is literally called Pure Stewed Lion's Head （清炖狮子头） in Chinese. Sweet and Sour Mandarin Fish is also a very classic dish. Its Chinese name is Squirrel Mandarin Fish （松鼠桂鱼）, because the dish is made into a shape of squirrel.

Sweet and Sour Mandarin Fish

Xiang Cuisine

Xiang cuisine mainly consists of local dishes from the Xiangjiang River area, Dongting Lake area and Western Hunan mountain area. Stemming from a "land of fish and rice", Xiang cuisine has a variety of local ingredients. It is characterized by its sour and hot flavor, fresh aroma and deep color. Xiang cuisine is also known for its large use of hot peppers. Hot peppers are often soaked in a sealed jar of vinegar, achieving a sour and spicy taste. Chopped chili pepper is also widely used in Xiang cuisine. Because of the humid climate in Hunan Province, people eat hot peppers to help resist dampness and coldness. Xiang cuisine favors cooking methods such as braising, steaming, smoking, stewing and stir-frying.

Dong'an Chicken (东安仔鸡), named by its material, the pullet, is one of the most famous Hunan dishes. There are other typical courses such as Steamed Multiple Preserved Hams (腊味合蒸), Steamed Fish Head with Chopped Pepper (剁椒鱼头), Maoist Braised Pork (毛氏红烧肉) and Hunan Spicy Chicken (麻辣子鸡).

Steamed Fish Head with Chopped Pepper

Hui Cuisine

Hui cuisine, also known as *hui bang* (徽帮) or Huizhou flavor, originates from the Huizhou Mansion in the Southern Song Dynasty, modern Huangshan city, Wuyuan county (婺源县) in Jiangxi province and Jixi county (绩溪县) in Anhui province. Hui cuisine mainly consists of the local cuisines in the south of Anhui province, areas along the Yangtze River and the Huaihe River. The formation and development of Hui cuisine are closely related to Anhui merchants whose high dietary requirements have contributed to the development of Anhui cuisine, making it more various and exquisite.

The highly distinctive characteristic of Anhui cuisine lies not only in the

elaborate choices of cooking materials, but also in the strict control of time in cooking. Besides stir-frying, frying, roasting and other cooking methods, Hui cuisine is well-known for steaming, braising and smoking. Inheriting the idea that medicine and food are originated from the same source in traditional Chinese medicine, Anhui cuisine gives top priority to the tonic food, which is another characteristic of Hui cuisine.

High up on the menu in Hui cuisine are Stewed Turtle with Ham (火腿炖甲鱼), Huangshan Braised Pigeon (黄山炖鸽), Steamed Rock Partridge (清蒸石鸡), Fried and Braised Moldy Tofu of Huizhou (徽州毛豆腐), etc.

Stewed Turtle with Ham

Zhe Cuisine

Zhe cuisine is composed of local dishes from cities of Hangzhou, Ningbo, Shaoxing and Wenzhou, among which Hangzhou cuisine enjoys the highest **prestige**.

Zhe cuisine wins its reputation for **exquisite** selection of seasonal ingredients, unique cooking techniques and **prominent** flavor of food. It is also known for fine cutting and elegant shaping of the dishes. Besides various fresh seafood and seasonal vegetables, the world famous Hangzhou Longjing tea leaves and Shaoxing aged wine (绍兴老酒) are frequently used in cooking.

The representative dishes of Zhe cuisine include West Lake Vinegar Fish (西湖醋鱼), Braised Bamboo Shoots (油焖春笋), Stir-fried Prawn with Longjing Tea (龙井虾仁), Braised Dongpo Pork (东坡肉) to name a few. Among them Dongpo Pork, a dish closely related to a **prestigious** poet Sushi[3], has become a must for visitors to Hangzhou.

Dongpo Pork

The eight regional cuisines, each with its own unique features, are the most influential and representative cuisines in

China. In pursuing the unity of color, aroma, taste and shape, each of the eight cuisines has achieved its unique style. Thus they are often described in the following vivid ways: Su and Zhe cuisine are often described as a graceful and elegant Southern beauty, Lu and Hui cuisine as a simple and robust Northern man, Min and Yue cuisine as a **refined** and romantic son of a wealthy family, and Chuan and Xiang cuisine as a talented and **versatile** celebrity. The eight major cuisines reflect the essence of Chinese cooking technique and represent the peak of Chinese food culture.

New words

cuisine [kwɪˈziːn] *n.* The cuisine of a country or district is the style of cooking that is characteristic of that place. 烹调风格

specialty [ˈspeʃəlti] *n.* A type of food that a person, restaurant, or area is well known for. （某人、餐馆或地方的）特色食品

renowned [rɪˈnaʊnd] *adj.* A person or place that is renowned for something, usually something good, is well known because of it. 有名的；有声望的

ingredient [ɪnˈɡriːdiənt] *n.* Ingredients are the things that are used to make something, especially all the different foods you use when you are cooking a particular dish. （尤指烹调用的）原料

course [kɔːs] *n.* A course is one part of a meal. 一道菜

delicacy [ˈdelɪkəsi] *n.* A delicacy is a rare or expensive food that is considered especially nice to eat. 美味佳肴

distinguished [dɪˈstɪŋɡwɪʃt] *adj.* If you describe a person or their work as distinguished, you mean that they have been very successful in their career and have a good reputation. 卓著的

ferment [fəˈment] *v.* If a food, drink, or other natural substance ferments, or if it is fermented, a chemical change takes place in it so that alcohol is produced. 使发酵；发酵

emigrate [ˈemɪɡreɪt] *v.* If you emigrate, you leave your own country to live in another country. 移居外国

flavoring [ˈfleɪvərɪŋ] *n*. Flavorings are substances added to food or drink to give it a particular taste. 调味品

integrate [ˈɪntɪgreɪt] *v*. If someone integrates into a social group, or is integrated into it, they behave in such a way that they become part of the group or are accepted into it. 使融入；结合在一起

seasoning [ˈsiːzənɪŋ] *n*. Seasoning is salt, pepper, or other spices added to food to improve its flavor. 佐料

prevalent [ˈprevələnt] *adj*. A condition, practice, or belief that is prevalent is common. 盛行的；普遍存在的

reputable [ˈrepjətəbl] *adj*. A reputable company or person is reliable and can be trusted. 声誉好的；可信赖的

delicate [ˈdelɪkɪt] *adj*. Something that is delicate is small and beautifully shaped. 精巧的；精美的

elaborate [ɪˈlæb(ə)rət] *adj*. Elaborate clothing or material is made with a lot of detailed artistic designs. （衣服、布料等）精致的；精巧的；制作精美的

prestige [presˈtiːʒ] *n*. If a person, a country, or an organization has prestige, they are admired and respected because of the position they hold. 威望

exquisite [ɪkˈskwɪzɪt] *adj*. Something that is exquisite is extremely beautiful or pleasant, especially in a delicate way. 精美的

prominent [ˈprɒmɪnənt] *adj*. Something that is prominent is very noticeable or is an important part of something else. 突出的

prestigious [preˈstɪdʒəs] *adj*. A prestigious institution, job, or activity is respected and admired by people. 有声望的

refined [rɪˈfaɪnd] *adj*. If you say that someone is refined, you mean that they are very polite and have good manners and good taste. 彬彬有礼的；举止文雅的

versatile [ˈvɜːsətaɪl] *adj*. If you say that a person is versatile, you approve of them because they have many different skills. 多才多艺的

Phrases and expressions

come into being　形成；产生；诞生
contribute to　促成；造成（某事发生）
give priority to　以…为主；强调；优先考虑

Cultural Notes

　　1. Li Bai（李白，701-762），also known as Li Taibai（李太白），was one of the most prominent figures in the flourishing of Chinese poetry in Tang Dynasty. It was said that the well-known Taibai Duck was named after Li Bai, who lived in Sichuan for nearly 20 years. Li Bai loved the dish so much that he presented it to the emperor, Tang Xuanzong, with the purpose of sharing his strategy of governing the country, but the emperor was only interested in the food and did not want to hear him talk about the affairs of the country. Li Bai was very disappointed and the meeting failed. Finally Li Bai was so frustrated that he decided to leave the court forever. After he left, the emperor named the dish Taibai Duck. Although Li Bai's political ambitions did not work in the imperial court, he left a fine and delicious dish and a good story for his people.

　　2. Fo tiao qiang（佛跳墙），the most famous dish in Min cuisine, also called Steamed Abalone with Shark's Fin and Fish Maw in Broth, is a variety of shark fin soup in Fujian cuisine. Since its creation during Qing Dynasty, the dish has been regarded as a Chinese delicacy known for its rich taste, usage of various high-quality ingredients and special manner of cooking. The dish's name is an allusion to the dish's ability to entice the vegetarian monks from their temples to partake in the meat-based dish. It is high in protein and calcium. It is also known as Buddha Jumps over the Wall.

　　3. Su Shi（苏轼，1037-1101），also known as Su Dongpo（苏东坡），

was a Chinese writer, poet, painter, calligrapher and a statesman of Song Dynasty. The famous Dongpo Pork was named after him. The legend has that people in Hangzhou all respected Su Dongpo very much. Knowing that Su was a fan of pork, they all presented him some pork on New Year's Day. Unable to decline their kindness, Su asked his family to braise all the pork with wine, soy sauce, ginger and sugar. Then the pork was sent to every household. People were very glad and called it affectionately Dongpo Pork. From then on, this dish has become one of the most traditional famous dishes of Hangzhou.

Exercises

1. Read the following statements and try to decide whether it is true or false according to your understanding.

_____ 1) The most famous traditional dish in Su cuisine is Sweet and Sour Carp.

_____ 2) The famous dish—*Fotiaoqiang*, steamed abalone with shark's fin and fish maw in broth is a representative delicacy in Min cuisine.

_____ 3) Chuan cuisine can be traced back to the ancient Ba Kingdom, the modern Chengdu city, and Shu Kingdom, the modern Chongqing city.

_____ 4) Steamed Fish Head with Chopped Pepper is one representative dish in Xiang cuisine.

_____ 5) Lu cuisine is often described as a celebrity, talented and versatile.

2. Fill in the blanks with the information you learn from the text.

1) The eight distinct regional cuisines in China are Lu, Chuan, _____, Min, Su, _____, Hui and Zhe.

2) The most characteristic cooking technique of Min Cuisine is cooking with _____.

3) _____ is well-known for its unique hot-spiciness which has an instantaneous _____ effect on the tongue.

4) Lu Cuisine mainly consists of Jinan cuisine, cuisine in Jiaodong area

and the Confucian-style cuisine, among which _____ is well-known for its well-chosen ingredients and elegant shaping of the dishes.

　　5) The famous Dongpo Pork belong to the _____ cuisine, while Taibai Duck is a dish in _____ cuisine.

3. Explain the following terms.

　　1) Sweet and Soor Carp
　　2) Hot Pot

4. Translate the following paragraph into English.

　　八大菜系各具特色，是中国最有影响力、最具代表性的菜系。八大菜系致力于追求色、香、味、形的统一，有人形象地把八大菜系描绘为：苏、浙菜好比优雅清丽的江南美女；鲁、微菜犹如健壮朴实的北方大汉；粤、闽菜宛如风流典雅的公子；川、湘菜就像多才多艺的名流雅士。八大菜系反映了中国烹饪技术的精华，代表了中国饮食文化的巅峰。

5. Critical thinking and discussion.

　　Some people think that the popularity of fast food like KFC and McDonald's may impact on Chinese food. Do you agree? Why or why not? Find some examples to illustrate your point of view.

Section B Food and Festivals

　　Many countries have traditional festivals, and China, with its long history, is no exception. Chinese festivals are celebrated in many different ways but there is no better way of enjoying these rich and colorful occasions with special tasty foods. All the festivals have their special foods, conveying some special **connotations**, mainly good wishes. For example, *jiaozi* and *niangao*, implying a wealthier year ahead and a better life in the future, are the traditional foods for the Spring Festival. Moon cake, **symbolizing** perfect life and family reunion, is a necessity for the Mid-autumn Festival. And there are some other traditional foods unique to certain festivals.

The Spring Festival

As one of the biggest traditional festivals in China, the Spring Festival lasts nearly a month. Ever since the eighth day of the twelfth month, people begin to prepare all kinds of foods for the New Year. There are so many foods expressing best wishes and high expectations for a future year.

Jiaozi is an **indispensable** food during the Spring Festival. Most families make and have *jiaozi* together on New Year's Eve. Making *jiaozi* conveys people's wishes to wrap the wealth inside while eating *jiaozi* means goodbye to the former year and welcome for the new one.

Jiaozi

The Chinese word "饺" has similar pronunciation with the word *jiao*(交), which means "cross". And *zi*(子) is the Chinese traditional time "23:00-1:00", the dividing point of the last year and the next year. In addition, the shape of *jiaozi* is like gold ingot (元宝) in ancient China. It is believed that eating *jiaozi* will help people to make a fortune. In order to make the festival more **joyous**, many families like to wrap some special things inside *jiaozi*, which conveys some special meanings. For example, sugar in *jiaozi* means happy and sweet life for the coming year. Peanut, also called **longevity** nut (长生果), means that people can live a longer life. Coins mean great fortune in the coming year. People in the northeast must eat *jiaozi* on the fifth day of the first month, which is known as **Break Five**[1]. Eating *jiaozi* expresses the meaning of keeping the bad luck at bay or getting rid of the evil.

Nowadays, *jiaozi* has become a very welcome staple food, especially in the north of China. It is frequently served on daily table and on one's birthday feast as well.

Niangao (年糕), another essential food for the Spring Festival, is very popular throughout China, especially in the south. It is made of the flour of glutinous rice and millet into the form of blocks or cakes, signifying silver and gold (金条银条), currency of ancient China. Eating *Niangao* during the

Spring Festival is believed to bring wealth in the coming year. *Niangao* also means "higher and higher, better and better one year after another" since *nian*（年）means "year" and *gao*（糕）has the same pronunciation with *gao*（高）in Chinese, which means "high" in English. *Niangao* in the north is usually sweet, while those in the south are both sweet and salty. In a word, the sweet and sticky *niangao* is a **token** for the New Year and represents a future year of happiness, wealth and good luck.

Niangao

The Lantern Festival

The Lantern Festival is celebrated on the fifteenth day of the first month. It marks the final day of celebrations for the Spring Festival. *Yuanxiao*（元宵）or *tangyuan*（汤圆）is the traditional food for Lantern Festival. It is a glutinous rice ball filled with sweet red bean paste, sesame paste, or peanut butter paste. Chinese people believe that the round shape of *yuanxiao* symbolizes family reunion since the pronunciation of *tangyuan* sounds similar to that of *tuanyuan*（团圆）in Chinese, which means togetherness and unity.

Yuanxiao has been **diversified** for thousands of years. As far as the skin is concerned, there are glutinous rice flour, sorghum flour, yellow rice flour and corn flour, etc. The fillings also range from the so-called osmanthus sugar（桂花糖）, hawthorn white sugar（山楂白糖）, bean paste, sesame and peanut to the assorted （什锦）fillings. The methods of production differ from north to south. In the north, *yuanxiao* is made by being rolled repeatedly in a square-bottomed bamboo basket, while in the south, it is made by making dough in one's

Tangyuan

hand. *Yuanxiao* can be as big as walnuts, or as small as yellow beans. It is usually served by being fried, steamed or boiled with soup. *Yuanxiao* has become a popular snack food in China all year round.

The Dragon Boat Festival

The Dragon Boat Festival is celebrated on the fifth day of the fifth month. For thousands of years, a notable part of celebrating the Dragon Boat Festival is making and eating *zongzi* with family members. This tradition is to **commemorate** the death of **Qu Yuan**[2] who drowned himself in the Miluo River on hearing his state was defeated. People dropped balls of sticky rice into the river so that the fish would eat them instead of Qu Yuan's body. This is said to be the origin of *zongzi*.

Zongzi

Zongzi is a pyramid-shaped dumpling made of glutinous rice. People usually wrap *zongzi* in leaves of bamboo, lotus or banana which give a special flavor to the sticky rice and fillings. Choices of fillings vary depending on regions. Northern people in China prefer sweet or dessert-styled *zongzi*, with bean paste, dates and nuts as fillings while Southern people prefer savory *zongzi*, with a variety of fillings including marinated pork belly, chicken, sausage and salted duck egg yolks. It is commonly accepted that the best known *zongzi* is from the cities of Huzhou and Jiaxing in Zhejiang Province. People in Sichuan province make a spicy *zongzi* with chili powder and **preserved** pork（腌肉）.

Zongzi is considered as a symbol of luck, as the pronunciation of *zong* is very similar to that of *zhong*（中）. This character has a positive connotation in Chinese, often used in such phrases *zhongjiang*（中奖）and *kejugaozhong*（科举高中）, which means winning a prize and scoring high in an exam, respectively.

Most people still keep the tradition of eating *zongzi* on the Dragon-Boat Festival. And the special delicacy has become so popular that it now can be enjoyed all the year round.

The Mid-Autumn Festival

The Mid-Autumn Festival is celebrated on the fifteenth day of the eighth month. Chinese people value this festival for its important meaning of reunion. Moon cake is the traditional food for this special occasion.

There is a folk tale about the origin of moon cakes. In order to **overthrow** the Mongol rule, Zhu Yuanzhang's adviser, Liu Bowen came up with a brilliant idea to unite the people with messages hidden in moon cakes. Word spread that a deadly plague was prevalent and the only way to prevent it was to eat special moon cakes,

Moon cake

which would help people **revive** instantly. This prompted the quick distribution of moon cakes to the Han people only. When eating the cakes, people found the message "Revolt on the fifteenth of the eighth month." Thus informed, they rose together on this day to overthrow Yuan Dynasty. Since then moon cakes have become an **integral** part of the Mid-Autumn Festival.

Most moon cakes consist of a thin, tender pastry skin with sweet and dense fillings, and may contain one whole salted egg yolk in their center as the symbol of the full moon. Popular fillings include lotus seed paste, sweet bean paste, date paste and five kernels（五仁）, among which, the five-kernel moon cakes are very classical with five commonly used nuts and seeds: walnuts, pumpkin seeds, sunflower seeds, peanuts, sesame seeds, or almonds. Traditional moon cakes have an imprint of the Chinese characters for "longevity" or "harmony", as well as the name of the bakery and the filling inside.

There are many kinds of moon cakes in China, such as the moon cakes of Beijing-style, Cantonese-style, Suzhou-style and Yunnan-style. Over time, both the skin and the fillings of moon cakes have diversified to cater to the changing taste. Traditional moon cakes are often sold alongside the new styles such as the Snowy Moon Cakes, the French-style moon cakes, to name a few.

Moon cakes are to Mid-Autumn Festival what **mince pies**[3] are to

Christmas. Today, it is **customary** for families to present them to their relatives and friends as gifts.

The *Laba* Festival

Laba (腊八) festival, celebrated on the eighth day of the twelfth month, is a very important festival in China, **signifying** the start of celebrations for the Chinese New Year. Families throughout China make a special porridge, called *laba* porridge, with rice, glutinous rice, millet, Chinese sorghum, red beans, peanuts,

Laba Porridge

dried lotus seeds, dried dates or other nuts. *Laba* porridge is also called the eight-treasure porridge, which was said to be introduced to China in Song Dynasty. It was said that large Buddhist temples would offer porridge to the poor to show their faith to Buddha. In Ming Dynasty, it became such a holy food that emperors would offer it to their officials during festivals. As it gained favor among the upper class, it quickly became popular throughout the country.

There is a touching story about the origin of *laba* porridge: When the founder of Buddhism, Sakyamuni (释迦牟尼) was on his way into the high mountains in quest of **enlightenment**, he grew tired and hungry. Thus he passed into unconsciousness by a river in India. A shepherdess (牧羊女) found him and fed him porridge made with rice and beans. Then Sakyamuni continued his journey. After six years of strict discipline, he finally realized his dream of full enlightenment on the eighth day of the twelfth month. Ever since, monks have prepared rice porridge on the eve and held a ceremony the following day, during which they offered porridge to Buddha. With time going on, the custom extended, especially in rural areas where farmers would pray for a plentiful harvest in this way. Gradually the tradition of having *laba* porridge has become very popular throughout China.

Traditionally, *laba* porridge is not only to be **sacrificed** to the ancestors and the gods, but also to the relatives, neighbors and even the passers-by. It is the best food that Chinese people use to gather everything around and express their love. It's a bowl of porridge for one billion and three hundred million Chinese people, for the world and for everything. In cold northern China, the eight-treasure porridge has become a welcome food for breakfast since it is served hot and full of nutrition.

Food to the festival is like water to the fish. These traditional foods have become an indispensable part for the celebrations of certain festivals. Some of them even become a cultural symbol of China, such as *jiaozi*, *zongzi* and *niangao*. Food is such an integral part of Chinese culture that to some extent, an understanding of Chinese food culture is an understanding of China itself. In order to construct a culturally powerful country, it is of great importance to promote the traditional food culture.

New words

connotation [ˌkɒnəˈteɪʃn] *n*. The connotations of a particular word or name are the ideas or qualities which it makes you think of. 内涵意义

symbolize [ˈsɪmbəlaɪz] *v*. If one thing symbolizes another, it is used or regarded as a symbol of it. 象征；代表

indispensable [ˌɪndɪˈspensəbl] *a*. If you say that someone or something is indispensable, you mean that they are absolutely essential and other people or things cannot function without them. 必不可少的

joyous [ˈdʒɔɪəs] *a*. Joyous means extremely happy. 快乐的

longevity [lɒnˈdʒevəti] *n*. Longevity is long life. 长寿；长命

token [ˈtəʊkən] *n*. A token is something that represents a feeling, fact, event etc. 象征；符号；标志

diversify [daɪˈvɜːsɪfaɪ] *v*. When an organization or person diversifies into other things, or diversifies their range of something, they increase the variety of things that they do or make. （使）多样化

commemorate [kəˈmeməreɪt] *v*. To commemorate an important event or person means to remember them by means of a special action, ceremony, or specially created object. 纪念

preserve [prɪˈzɜːv] *v.* If you preserve food, you treat it in order to prevent it from decaying so that you can store it for a long time. 保藏；腌制

overthrow [ˌəʊvəˈθrəʊ] *v.* When a government or leader is overthrown, they are removed from power by force. 颠覆

revive [rɪˈvaɪv] *v.* If you revive someone who has fainted or if they revive, they become conscious again. 使苏醒；苏醒

integral [ˈɪntɪɡrəl] *a.* Something that is an integral part of something is an essential part of that thing. 构成整体所必需的

customary [ˈkʌstəməri] *a.* Customary is used to describe things that people usually do in a particular society or in particular circumstances. 通常的

signify [ˈsɪɡnɪfaɪ] *v.* If an event, a sign, or a symbol signifies something, it is a sign of that thing or represents that thing. 表示；意味着；意思是

enlightenment [ɪnˈlaɪtnmənt] *n.* Enlightenment means the act of enlightening or the state of being enlightened. 启发；教化

sacrifice [ˈsækrɪfaɪs] *v.* To sacrifice an animal or person means to kill them in a special religious ceremony as an offering to a god. 献祭

Phrases and expressions

make a fortune　发财；赚大钱
keep sth at bay　阻止某事发生；使某物不能接近
range from sth to sth　包括（各种不同的人或物）
a variety of sth　各种各样的某事物

Cultural Notes

1. Break Five（破五）, a traditional Chinese custom, celebrated on the fifth day of the first lunar month. It is said that God of Fortune's birthday falls on this day also. People in the North usually celebrate this day with a large banquet. According to the traditions, nobody visits friends or relatives as it would bring bad omen. They stay at home to worship the God of Fortune. This day is also commonly known as the Festival of Po Wu, literally breaking five. According to custom, it is believed that many New Year taboos can be broken on this day. Eating *jiaozi* on the day conveys the meaning of keeping the bad luck at bay or getting rid of the evil.

2. Qu Yuan（屈原，340-270 BC）, a great poet in ancient China, was extremely patriotic. After the capital of Chu was seized by the Qin army, angry and desperate, Qu Yuan drowned himself in Miluo River. Legend has it that the day Qu Yuan drowned himself was on May 5th on China's lunar calendar. To commemorate Qu Yuan, on that day each year, people hold dragon-boat racing and eat *zongzi*. Holding dragon boat racing is to scare fish away; feeding fish with *zongzi* is to stop them from eating Qu Yuan's body. Adopting the form of folk ballad, Qu Yuan created a new poetic form called "the songs of Chu" in Chu dialect. Qu Yuan wrote many wonderful poems, among which *Li Sao* is most famous. His works are characterized by beautiful language and fancy imagination.

3. Mince pie（百果馅饼）, a sweet pie of British origin, filled with a mixture of dried fruits and spices called "mincemeat", is traditionally served during the Christmas season in the English-speaking world, excluding the USA. Its ingredients are traceable to the 13th century, when returning European crusaders（十字军）brought with them Middle Eastern recipes containing meats, fruits and spices.

Chapter 10　Chinese Food Culture

Exercises ...

1. Read the following statements and try to decide whether it is true or false according to your understanding.

____ 1) *Jiaozi* is an important staple food for the people in the south of China.

____ 2) People in the northeast must eat *jiaozi* on the fifth day of the first month, also known as "Break Five".

____ 3) Southern people in China prefer sweet or dessert-styled *Zongzi*, with bean paste, dates and nuts as fillings.

____ 4) The five-kernel moon cakes are very classical with five commonly used nuts and seeds: walnuts, pumpkin seeds, sunflower seeds, peanuts, sesame seeds, or almonds.

____ 5) The *Laba* Porridge is also called the eight-treasure porridge, which was first introduced to China in Song Dynasty.

2. Fill in the blanks with the information you learn from the text.

1) The Spring Festival is the Chinese lunar New Year's holiday featuring _____ and _____ as the traditional food.

2) Moon cake, symbolizing _____ and _____, is indispensable for the Mid-Autumn Festival.

3) The traditional food for Lantern Festival is *yuanxiao*, which is a glutinous rice ball typically filled with sweet red bean paste, _____ or peanut butter.

4) The most famous *zongzi* in China are those from Zhejiang's _____ and _____, and these have become the national standard

5) The _____ is one of the most traditional festivals in China. And Chinese people value this festival for its important meaning of reunion.

3. Explain the following terms.

1) *Niangao*

2) Eight-treasure Porridge

4. Translate the following paragraph into English.

中国人非常重视中秋节，因为中秋节象征着家庭团圆。月饼是中秋节的经典食品。传统月饼上面有"长寿"或者"和谐"字样，同样印有月饼商的名号和月饼的馅料。中国月饼种类繁多，有京式月饼、广式月饼、苏式月饼、滇式月饼等。随着时间的推移，月饼的饼皮和馅料都日渐丰富多样化，以迎合人们不同的口味需求。传统月饼与新式月饼（如冰皮月饼、法式月饼等）并存。

5. Critical thinking and discussion.

The majority of Chinese people cherish the traditional festivals and keep the habit of preparing some typical foods for the special festivals, while some youngsters have no idea about how to make some traditional foods. Do you think it is necessary to teach the young people to make some traditional foods? Why or why not?

课后练习题答案

Chapter 1

Section A

1. 1) F 2) T 3) F 4) F 5) T
2. 1) Pangu, Quafu 2) *lei*, *si* 3) extraordinary running ability
   4) Emperor Yan 5) *Jingwei Tries to Fill the Sea*
3. 1) Pangu was a legendary creation of the world. He was said to wake up inside an egg full of burry and cold substances. He has been sleeping for 18,000 years till one day he broke it with a bang. Then light and transparent substances ascended and turned into heaven while those heavy and murky things gradually deposited into earth. Tens of thousands of years passed, the heaven and the earth stretched beyond measurement and then stopped changing. The world finally came into being.

    2) *Jingwei Tries to Fill the Sea* tells the sad story of the youngest daughter of Emperor Yan. It was said she drowned herself when playing in the east sea. She turned into a bird and mourned herself sadly in the sound "jing-wei, jing-wei". Hence she was named "Jingwei". She hated the sea for taking her life ruthlessly and decided to fill the sea for a revenge. Every day, she flew to and fro between the mountain and the sea, carrying twigs and pebbles and dropping them into the sea. From the myth comes the Chinese idiom "*Jingwei tries to fill the sea*".

4. Myth is the product of the beginning time of human society. It reflected the ancient people's magnificent interpretation and wild imagination of natural phenomena. All the natural forces were vividly visualized and personified in their mind. Then they invented many stories about gods and passed them around verbally based on the heroic figures in real life. That's the origin of myth. The reading of myths can help people keep well-informed of the past,

well-adapted to the present and even well-prepared for the future.

5. （略）

Section B

1. 1) F 2) F 3) T 4) T 5) T
2. 1) Mandate of Heaven 2) heaven, spirit tablet 3) scripture chanting
   4) *chun ji*, *zha ji* 5) Thunder and Lightening
3. 1) Heaven, or Haotian God, is the most active, influential and overawing god in Chinese ancient life. Heaven sacrifice shares uttermost importance whether to the ancient emperors or the commons. Its real purpose is to please Heaven in order to get blessings with respectful attitude, bending behaviors and rich food offerings. Heaven sacrifice has developed a series complicated and strict procedures. It's been documented and listed as the most significant matter of China. The Emperors in different dynasties have paid priority to Heaven sacrifice.

   2) Ancestor worship is a ritual practice on the belief that deceased family members have a continued existence. It begins at the deceased people's funeral where a home altar is set up by their family members. The living would make offerings and observe a simple living style in order to show their respect, remembrance and sorrow. After the funeral, the corpse of the deceased would be buried with sacrifices, typically things he or she was thought to need in the afterlife. This was done as a symbolic demonstration of being filial. The goals of ancestor sacrifice, regardless of religious beliefs, are to display respect, provide comfort for the deceased in another world, protect their offspring against evil spirits and ensure the ways of the deceased's soul into the afterlife.

4. Sacrifice is a part of Chinese rituals and the important segment of Ruist ceremony. Sacrifice means to respect gods, seek blessings from them and worship ancestors. In primitive times, people were shocked at the natural forces. They assumed that some supernatural forces could dominate the fate of human. And sacrifice was a ceremony that enabled people to communicate with those forces. Sacrificial activities were relatively simple and unorganized at first. People made the image of gods with bamboo, wood or clay, or draw the iconic pictures of the sun, the moon, the stars and wild animals on

rocks. Then they laid out food and other tributes before gods and spirits, sang and danced to them to show their respect and ask for protection.

5. （略）

Chapter 2

Section A

1. 1) F 2) T 3) F 4) F 5) T
2. 1) Duke Xiao of Qin 2) Northern Wei 3) the local governments
   4) state-owned land 5) Fengyang
3. 1) The well-field system was a Chinese land distribution method in effect between the late Western Zhou Dynasty to around the end of the Warring States period. Its name comes from Chinese character *jing* （井）, which means "well" and looks like the ♯ symbol. This character represents the theoretical appearance of land division: a unit of land was divided into nine identically-sized sections, the eight outer sections were privately cultivated by serfs and the center section was communally cultivated on behalf of the land-owning aristocrats.

   2) People's Livelihood Principle of Sun Yat-Sen is the equalization of landownership and the restriction of capitalism with a view to uprooting the ills of capitalism. The equality in land properties was one of the important contents. The aim was to realize the transformation of the old land system to "state-owned land", and to avoid the social problems of hostility between the rich and the poor after the rise of land price with the development of business. Under this principle, the Agrarian Revolution put into effect the principle of "land to the tiller" and would then "turn the land over to the private ownership of the peasants." The economic rent should be socialized and shared by everybody in the society, not monopolized by the landlords.

4. Wang Anshi was a famous thinker, politician, writer and reformer in Northern Song Dynasty. During the reign of Emperor Shenzong of Song, Wang Anshi carried out the new policies in the whole nation and launched the campaign of reform in a very large scale, involving all the aspects such as politics, finance, military and society and so on. Its purpose was to

enrich the country and increase its military force so as to change from being poor and weak. While, the reform was bitterly opposed by the big landowners and failed ultimately because of affecting their interests. Though, the Reforms of Wang Anshi still played a very positive role in the social development of the Northern Song at that time.

5. （略）

Section B

1. 1) T 2) F 3) T 4) F 5) T
2. 1) compartment method 2) alcohol making 3) *Chen Fu Nongshu*
   4) farm tools 5) famine relief
3. 1) *Qimin Yaoshu* was written by Jia Sixie. The book mainly recorded the agricultural production in the lower reaches of the Yellow River basin. It covers a wide range of topics, detailing agronomy, horticulture, forestry, sericulture, animal husbandry, breeding, brewing, cooking, reserves, and disaster relief method, the processing of agricultural and sideline products such as brewing and food processing etc. Besides, the book described different methods of plant propagation like striking, layering, division or propping. The book is worth being an "ancient Chinese agricultural encyclopedia".

   2) The book *Nongzheng Quanshu* was written by Xu Guangqi and actually not completed during his lifetime, but left to his disciple Chen Zilong who amended the text and rearranged it. The book completely reviews all aspects of farming in ancient China and shows the level that agriculture had reached in the seventeenth century. The book remarkably discusses land reclamation, water conservancy and famine relief, which are rare in other agricultural books. What is more important, the book features Xu Guangqi's thoughts of agriculture politics which advocates that agriculture development is the root of a rich country. This is the basic idea that runs through the book.
4. Jia Sixie was a famous Chinese ancient agronomist. He stressed that agricultural production must follow the laws of nature and must not disobey the farming seasons. And he emphasized that the agricultural production techniques and tools also needed to be reformed. He wrote an agricultural

books, *Qimin Yaoshu*. This book summarizes the production experience accumulated by working people in the north of our country. And it mainly introduces the production method of agriculture, forestry, animal husbandry, sideline and fishery. It is usually viewed as the most complete encyclopedia of agriculture in our country, and it is an important resource to study the agricultural production in ancient China.

5. （略）

Chapter 3

Section A

1. 1) T 2) T 3) T 4) F 5) F
2. 1) northern hemisphere 2) Insects Awakening 3) Grain in Ear
   4) Winter Solstice 5) *Huainanzi*
3. 1) The traditional Chinese calendar divides the year into 24 solar terms based on observation of the sun's annual movement. The names of 24 terms were chosen according to changes in temperature, precipitation and other natural phenomena. For instance, some of the terms reflect the change of seasons, such as Start of Spring and Start of Summer; some indicate the changes of height of the sun, such as Spring Equinox and Autumn Equinox; some reflect the changes of temperature, such as Major Heat and Limit of Heat and some show changes of precipitation like Clear and Bright and The Rains. Originated from the Yellow River reaches in ancient times, the use of the 24 solar terms has since been widely adopted by other places across China.

   2) Clear and Bright, or "Qing Ming" in Chinese, is an important solar term. The words "clear" and "bright" describe the clear air and bright scene in spring during this period. The first day of Clear and Bright becomes a traditional Chinese festival, Qingming Festival. Clear and Bright is a period in which Chinese people honor nature and the ancestors. it is a not only period for commemorating the deceased, but also time for people to go out and enjoy nature.

4. From the names of the 24 solar terms, we can see that their division follows the changing seasons and climates. Among them, Start of Spring, Start of

Summer, Start of Autumn, Start of Winter, Spring Equinox, Autumn Equinox, Summer Solstice and Winter Solstice are related to 4 seasons: spring, summer, autumn and winter. Spring Equinox, Autumn Equinox, Summer Solstice and Winter Solstice fall under the discipline of astronomy and are used to indicate the changes in the height of the sun. Start of Spring, Start of Summer, Start of Autumn and Start of Winter show the starts of the four seasons. China has a large territory and the climates in different regions vary greatly, thus, the changes of the four seasons vary greatly.

5. (略)

Section B

1. 1) T 2) F 3) T 4) T 5) F
2. 1) Goumang 2) cattle 3) Grain in Ear 4) White Dew 5) wheat planting, resist cold
3. 1) Around Autumn Equinox, farmers all over the country begin to experience *sanmang* which means they are extremely busy harvesting, plowing, and sowing. On the one hand, farmers are on the wing harvesting cotton and late wheat. On the other hand, they are busy plowing and selecting the seeds of wheat, barley and broad beans to prepare for sowing. During this period, the autumn rain is still the problem facing the farmers. Therefore, they should harvest the crops as early as possible in case of the continuous rain and the coming frost. And they should sow the winter crops as early as possible so as to make full use of the heat resources before winter and to breed strong seedlings for overwintering, which may lay a foundation for high yield in the coming year.

   2) Plum Rains, often occurring during June and July, refer to the long period of continuous rainy or cloudy weather. This happens to be the time for plums to ripen, which explains the origin of its name. During Plum Rain season, high humidity and high temperatures make clothes go moldy which is why it is also called "mold rain season". Plum Rains is a good period for growing rice, vegetables and fruits.
4. The 24 solar terms are established in ancient China with the purpose to

guide the farming activities. At the early stage of farming, people have found the close relationship between agricultural activities and the season. Till Qing Dynasty, a complete 24 solar term is established through continuous observation and summary. Some of the solar terms are named even directly after the farming term, such as "Xiao man" and "Mang zhong". The 24 solar terms objectively reflect the changes of the seasons and climate conditions of reaches of the Yellow River, which is the birthplace of the Chinese. And it is a concrete manifestation of the ancient Chinese agricultural civilization.

5. （略）

Section C

1. 1) F 2) T 3) F 4) T 5) F
2. 1) *yaochun* 2) light, greasy 3) *jiaozi*, noodles 4) autumn weariness
   5) egg
3. 1) The Cold Food Festival or Hanshi Festival is a traditional Chinese holiday celebrated for one day before the Clear and Bright. Legend has it that Prince Chong'er of the State of Jin during the Spring and Autumn period, endured many hardships while he was exiled from his home state. Once, in order to help the Prince who was tormented by hunger, Jie Zitui cut off the flesh from his thigh and offered it to the Prince for sustenance. When Chong'er became Duke Wen of Jin, however, he did not reward him, instead killing Jie and his mother in a fire. Later, Duke Wen, filled with remorse, ordered that using fire on the anniversary of Jie's death was forbidden and all food was to be consumed cold.

   2) On the first day of the Beginning of Autumn, usually people will weigh themselves and compare their weights to what it was at the Beginning of Summer. If one has lost weight during the summer, then at the beginning of autumn, he or she will have a variety of delicious food, especially meat, to regain the lost weight in summer, called *tieqiubiao* in Chinese.

4. From the perspective of health preservation, the changes of human bodies and the diseases are closely connected with the 24 solar terms. Therefore,

people should adjust their diet according to the changes of the seasons. People should choose different food and take different activities based on the characteristics of each solar term in order to achieve the purpose of longevity and health.

5. （略）

Chapter 4

Section A

1. 1) T 2) F 3) T 4) T 5) F
2. 1) gathering; producing; New Stone Age 2) millet 3) population migration
   4) the Yellow River Valley; Yangtze River Valley 5) cotton
3. 1) Five Grains is an umbrella term that covers the most important food crops cultivated by Chinese, such as millet, broom-corn millet, rice, wheat and soybean. Chinese legends attribute the Five Grains to Shennong, the legendary Emperor of China. They are considered as the most representative food crops cultivated in ancient China.

   2) *Sheji*: *Ji* (稷), in Chinese, refers to millet, one of the grains planted in China. Later, some sacred meanings were added to it. It became the embodiment of God of Grains. Another god that was in charge of harvest was the God of Land, which was called *she* (社) in Chinese. Since Zhou Dynasty in China, *sheji* became a fixed term symbolizing China the state, from which we can also get a glimpse of the importance of millet in ancient China.
4. In ancient times, there were many different versions of "Five Grains". The two major ones are as follows: one refers to rice, broomcorn millet, millet, wheat and soybean. Another refers to hemp, broomcorn millet, millet, wheat and soybean. The difference lies in that the former version includes hemp instead of rice, which appears in the latter. As to why there is no rice included in Five Grains at beginning, it can be attributed to the fact that at that time the economic and cultural center was in the Yellow River valley, where the production of rice was scarce since it was mainly produced in the south. With socio-economic development and agricultural production improvement, the concept of five cereals has been constantly

extended. Now the so-called Five Grains become the general name of food crops, or refer to food crops in general.

5. （略）

Section B

1. 1) T 2) F 3) F 4) F 5) T
2. 1) plough 2) plowshare, mouldboard 3) directions 4) depth; rows
   5) Shadoof
3. 1) Plough refers to the farming tools used for soil preparation. It was invented in Spring and Autumn period when iron smelting technology and animal power started to be in use. In comparison with other tools, plough represented a significant breakthrough in the invention of farm tools, as it could be moved continuously while the movement of other tools was intermittent or on and off. In the history of primitive agriculture, the invention of plough created the technological conditions for tremendous increase in productivity. The most representative ploughs were moldboard plough in Western Han, which combined the two operations of turning over and crushing the earth. The other one is the curved-shaft plough which could turn both left and right, even turn around while working, which made it more flexible. The design of modern ploughs are based on the same principles.

   2) Seedplough is a farming tool for sowing. It was invented in Warring States Period. The machine consists of a frame, a hopper and a tube with a spade that can break the soil. The tube can reach a specific depth under the ground, where seeds are dropped and covered at once. At the beginning, it was single-footed and later dual-footed. During Han Dynasty, an officer named Zhaoguo invented a tri-foot seed plough based on the previous ones, and the new tool can seed in three rows at the same time, thus greatly prompted the agriculture. The seed plough allowed farmers to sow seeds in well-spaced rows at specific depths and at a specific seeding rate. The use of a seed drill also facilitated weed control. The technology of making multi-tube seed drill was 1,100 years ahead of Europe.

4. In comparison with previous tools like *leisi*（耒耜）, shovel and hoe,

plough represented a significant breakthrough in the invention of farm tools, as it could be moved continuously while the movement of other tools was intermittent or on and off. In the history of primitive agriculture, the invention of plough created the technological conditions for tremendous increase in productivity.

5. （略）

Chapter 5

Section A

1. 1) T 2) F 3) T 4) F 5) F
2. 1) barrier lakes 2) building thick walls with earth 3) Shun 4) Xia Dynasty
   5) respect laws of nature, changing circumstances
3. 1) The story, *Da yu zhi shui*, is widely known in myth and legends of China. Yu is the hero of controlling flood along the Yellow River Basin. It is said that in the reign of Yao and Shun, the overrunning flood was mischievous, so Yao assigned Gun to regulate it. Gun failed with embankment and was killed by Shun. Then, Shun recruited Yu, Gun's son, to continue the work. Spending a backbreaking thirteen years and bypassing his house three times but never going in, Yu dredged new river channels as outlets, guiding water to river and river to sea. This not only eliminated the flood, but contributed to the development of agriculture. Yu was revered as Yu the Great for his achievements and selected to succeed to the throne by Shun. Yu's water control principle has had far-reaching impacts on those who harness rivers later.

   2) Yu created a vast network of channels, allowing the flood waters to flow along river courses into the sea. Yu's water control principle "dredge and channel rivers to drain the flood waters" has had far-reaching impacts on generations of people who harness rivers. According to the principle, human efforts are important in conquering nature, but it's more important for people to respect laws of nature and then take actions in light of changing circumstances. And this concept or value has been gradually adopted to address problems in actual reality and incorporated into Chinese culture.

4. The Yellow River is the cradle of the Chinese nation. She has fostered generations of Chinese but she has also brought troubles with her floods. The diligent Chinese people have been struggling to control the water of the river for generations. Yu was a famous expert on water regulation in China more than 4,000 years ago. Da Yu inherited the experience from his father, who once controlled the floods, and after plenty of investigations and researches, he found the cause of the floods. Da Yu worked very hard. He had been working on controlling the floods for 13 years constantly. During this period he did not go back to his home, even stopping by without going into his house for 3 times.

5. （略）

Section B

1. 1) F   2) T   3) F   4) T   5) T
2. 1) the Dujiangyan Irrigation System   2) the Zhengguo Canal, the Lingqu Canal
   3) Fish Mouth Levee the Min River   4) Mountain Yulei, the Baopingkou
   5) *tian fu zhi guo*, a land of abundance
3. 1) The Yuzui or Fish Mouth Levee is the key part of the Dujiangyan system, named for its cone-shaped head that resembles the mouth of a fish. It is an artificial levee that divides the water into inner and outer streams. The inner stream bed is deep and narrow, while the outer stream bed is relatively shallow but wide. This special structure ensures that the inner stream carries about 60% of the river's flow into the irrigation system during dry season. While during flood, this amount decreases to 20% to protect the people from flooding. The outer stream drains away the rest, flushing out much of the silt and sediment.

   2) The Baopingkou or Bottle-Neck Channel, which Li Bing gouged through the Yulei Mountain, is the final part of the Dujiangyan system. The channel distributes the water to the farmlands in the Chengdu Plain. As a check gate, the narrow entrance creates the whirlpool flow that carries away the excess water from Flying Sand Weir, effectively preventing flooding. Interestingly, the name Baopingkou derives from the narrow entrance.

4. The Dujiangyan Irrigation System was constructed by Li Bing 256 BC over

2,200 years ago. It is the oldest and only project retained in the world so far and characterized in its successful control of water without any blocking dam. The Chengdu plain looks like an open paper fan slanting to the southeast, while Dujiangyan is just at the top of the fan-shaped plain over 700 meters above sea level. Li Bing made rational use of this natural sloping terrain and built it without destroying natural resources. The Dujiangyan system is mainly composed of three parts: fish mouth levee, flying sand weir and bottle-neck channel. After the completion of Dujiangyan, the Chengdu plain has a vast field of fertile land, so that Chengdu has the reputation of "the land of abundance". Dujiangyan is not only a famous ancient water conservancy project, but also a famous scenic spot. In 2000, it was listed as the world heritage list by UNESCO as a cultural heritage.

5. （略）

Section C

1. 1) T 2) F 3) T 4) T 5) T
2. 1) the Cradle of Chinese Civilization, China's Sorrow   2) Jia Rang
   3) Ming Dynasty, main channel, two layers of dikes   4) Jin Fu, Chen Huang
   5) Three-North Shelter Forest Program
3. 1) Pan Jixun, Jin Fu and Chen Huang tried to control floods by constructing the artificial dams to regulate the water flow. Their measures embody Confucianism, which is of discipline and order imposed upon nature. Pan Jixun, a water conservancy expert, presided over governing the Yellow River. In the long-term practice, he learned from the predecessors and summed up Chinese people's experience in water control. Consequently, he regulated the levee system again, blocked many branches of the river and made the river flow in a main channel. What's more, he built two layers of dikes to confine floods. Most importantly, he proposed the strategy of flushing sediment with converging flows by narrowing the channel, which had a great impact on later generations to deal with the river control. His philosophy of flood prevention and river regulation had significantly changed from diversion to embankment ever Since. In the Qing Dynasty, many talented administrators of river engineering had done much in

river regulation such as Jin Fu and Chen Huang. Jin and Chen applied Pan's theory and practiced "converging flow with narrow channel and scouring sand with high flow rate". In other words, the river management represented a tremendous interference with nature, in contrasts with the Taoism solution.

2) The Taoism solution of flood control is allowing the river a more "natural" course within lighter constraints. The flooding projects before Ming Dynasty reflected Taoism of river management. In other words, the management showed more respects for it. For example, in Western Han Dynasty, an engineer Jia Rang proposed channelizing the Yellow River, improving its rate of flow to the sea but at the cost of fields and towns. Another plan was to divert the excess water so that floods would be mitigated. Wang Jing in Eastern Han Dynasty also stabilized the Yellow River with many diversion channels and constructed water gates.

4. The Yellow River ranks the third longest in Asia and the sixth longest in the world. The word "yellow" describes the muddy water of the river. The Yellow River, one of several rivers for China to live on, originates from Qinghai, flows through nine provinces, and finally pours into the Bohai Sea. The river basin is not only the birthplace of ancient Chinese civilization, but also the most prosperous region in the early history of China. However, due to the frequent devastating floods, it has caused many disasters. In the past few decades, the government has taken various measures to prevent disasters.

5. （略）

Chapter 6

Section A

1. 1) T 2) F 3) F 4) F 5) T
2. 1) the Spring Festival 2) *Zongzi* 3) Double Seventh Festival
   4) the Reunion Festival 5) moon cakes
3. 1) The dragon boat race is the grandest activity on the Dragon Boat Festival, prevailing in the south of China where water in rivers and

lakes is plentiful. The race originated from the attempt to rescue Qu Yuan who committed suicide by throwing himself into Miluo river. Today, people still have dragon boat races to commemorate the attempt to save Qu Yuan on the fifth day of the fifth lunar month every year.

2) Like eating *zongzi* on the Dragon Boat Festival and eating tangyuan on the Lantern Festival, eating moon cakes is a traditional Chinese custom to celebrate the Mid-Autumn Festival. On the night of the festival, the family will get together, eat moon cakes, enjoy the bright moon and chat leisurely. The moon cake is usually round, symbolizing reunion.

4. The Spring Festival is the grandest and the most distinctive traditional festival of Chinese folk. Generally referred to New Year's Eve and January 1st, it is the first day of the year, also known as the lunar year, commonly known as "New year." During the Spring Festival, Han people and many ethnic minorities in China hold various activities to celebrate it, which feature the events such as worshiping the gods and Buddhas, paying tribute to ancestors, removing the old, meeting the new and praying for the good year and so on. The activities are rich and colorful, with a strong national characteristics.

5. （略）

Section B

1. 1) T 2) F 3) T 4) T 5) F
2. 1) wedding night 2) matchmaker 3) the wedding ceremony 4) coffin
   5) a piece of black cloth
3. 1) The six important rituals about traditional wedding custom in China are known as *liuli*. They include proposing marriage, matching birth dates, submitting engagement gifts, presenting wedding gifts, selecting a wedding date and holding a ceremony. Nowadays, the six etiquettes still apply in most of the rural areas of our country.

   2) Holding the funeral wake refers to the period in which the deceased's family needs to stay awake around the deceased, which is a way for loved ones to show filial piety and loyalty to the deceased. The funeral wakes range from three to seven days, enabling relatives and friends to

pay their last respect to the deceased.
4. Performing the marriage ceremony, also known as "bowing to Heaven and Earth", is a very important ceremony in wedding. The marriage custom was very popular after Song Dynasty. After performing the marriage ceremony, the woman will become a member of the male family. During the ceremony, the person presiding over the wedding will say loudly: "First, bow to heaven and earth; second, bow to your parents; third, bow to husband and wife, and then enter the bridal chamber, please." In fact, the first bow represents the worship of the gods of heaven and earth. The second bow is the embodiment of filial piety. And the third bow represents the respect and affection of husband and wife.
5. （略）

Section C

1. 1) T 2) F 3) F 4) T 5) F
2. 1) commercial activity 2) enjoyment 3) tasty snacks 4) Ditan Temple Fair
   5) God of Earth
3. 1) There are various artistic folk performances on display in the temple fairs. The yangge, stilts, land boat, lion dance, drum dancing, gong and drum beating, acrobatics, magic shows, operas and cross-talk shows perform again and again for the enjoyment of the visitors. In fact, traditional shows such as cross talks, gongfu shows and lion dances have become the major features of the fair. The artistic folk performances at temple fairs are quite appealing to visitors.
   2) Ditan Temple Fair is one of the most famous temple fairs in China. Located in Beijing, this temple fair, during the Spring Festival, is one of the most important activities and a traditional cultural event that features all kinds of Chinese folk art. The fair, during the Spring Festival, has lots of games to play, food to eat, performances and lots of people. This temple fair is also popular among foreigners who have access to Chinese traditional culture by appreciating craftsmanship and artwork displayed by local artisans at the fair.
4. The temple fair is a religious custom of Chinese folk, generally held in the Lunar New Year, and Lantern Festival Festival. It is also one of the forms

of trade in Chinese bazaars, and its formation and development are related to the religious activities of the local temples. Held on a temple festival or on a fixed date in and around the temple, temple fair offers worship, entertainment and shopping. The temple fair is popular in many parts of the country. It is also a traditional folk activity in China.

5. (略)

Chapter 7
Section A

1. 1) F 2) T 3) F 4) T 5) F
2. 1) weave 2) flexible 3) dumping 4) Double-sided 5) the Silk Road
3. 1) In Qing Dynasty, the four famous embroideries (Su, Xiang, Shu and Yue) enjoyed a high reputation. Among them, Su Embroidery was well known for the exquisite designs, elegant colors and unique patterns. The product surface was flat, the rim neat, the stitch fine, the lines dense, and the color harmonious. Its products fell into three major categories: costumes, decorations for halls and crafts for daily use. Double-sided embroidery is an excellent representative. The typical work is an embroidered cat with bright eyes and fluffy hair, looking vivid and lifelike.

   2) Matouniang is about a love story between a girl and her white horse. Once upon a time, a girl in today's Sichuan Province badly missed her father serving in the army from afar. She promised her white horse a marriage if it could take her father home. However, she ate her words after the horse succeeded in bringing her father back. Heartbroken by her betrayal, the horse whickered and stopped feeding. His unusual behavior drew the father's attention, so he questioned his daughter and got the whole story. Frightened by the horse's strong determination to marry his daughter, he killed and skinned it, and hung the horsehide in the yard. One day, the horsehide wrapped up the girl and fled away. Shortly afterwards they were found to have transformed into a silkworm with horse head and human body in the mulberry wood. That is why the white worm was named Matouniang.

4. China is home to silk. In Shang Dynasty, the weaving skills reached a relatively high level. Complicated loom and figured silk emerged. Since the Western Han Dynasty, silk goods starting from Chang'an were continuously transported to foreign countries on a large scale through the Silk Road. It is soft, lightweight and breathable, but easily fades, wrinkles and stains. Silk is popular for its extraordinary quality, superb craftsmanship and exquisite design and color. It has reflected the collective wisdom of ancient Chinese craftsmen and their unremitting pursuit of beauty.

5. （略）

Section B

1. 1) T 2) T 3) T 4) T 5) F

2. 1) Terra Cotta Army 2) *wucai*/five-color porcelain, *doucai*/ contending colors porcelain 3) Ru, Guan, Ge, Jun, Ding 4) Yuan 5) Famille-rose

3. 1) *Tang Sancai* was the most representative pottery in Tang Dynasty, which was invented on the basis of lead-glaze pottery of Han Dynasty. Its glaze mainly featured the three colors of yellow, green and white, hence the name *Sancai* or tri-color. To make *Tang Sancai* Pottery, the glaze colors of yellow, green, blue, white, auburn and black were also applied in an interlaced way on the same ware. *Sancai* pottery was made not only in such traditional forms as bowls and vases, but also in the more exotic guises of camels and Central Asian travelers, testifying to the cultural influence of the Silk Road. Tang Sancai pottery reached its peak in the High Tang period and fell off after the late-Tang period. Even though the glory was short lived, its impact on the development of colored pottery and porcelain was great. *Tang Sancai* remains a most favored handicraft works for people today.

2) Jingdezhen is known as the "Porcelain Capital", and its original name was Xinpin and Changnan. Emperor Jingde decreed all the pieces made for court to be marked "made in the Jingde period" and thus the city changed its name to Jingdezhen. The city boasts a long history of porcelain making. Historical records show that porcelain making in Jingdezhen dates back to Han Dynasty. Its most productive period,

however, began during Song Dynasty. In Yuan Dynasty, its position as porcelain making center was consolidated, for the government office Fuliang Porcelain Bureau was set up there. In Ming and Qing Dynasties, the ceramics industry in Jingdezhen flourished, producing high-quality ware on a vast scale for export and for the imperial household as well. The most famous porcelain from Jingdezhen is the blue and white porcelain, which was started in Yuan dynasty. The porcelain produced there was so exquisite that it was as white as jade, as bright as a mirror, as thin as paper, with a sound as clear as a bell. Today, Jingdezhen remains a national center for porcelain production in China.

4. Porcelain is made from well-chosen porcelain clay or pottery stone through technological processes. Although porcelain is developed from pottery, the two are different in raw material, glaze and firing temperature. Compared with pottery, porcelain has tougher texture, more transparent body and finer luster. Porcelain promotes economic and cultural exchanges between China and the outside world, and also profoundly influences the traditional culture and lifestyle of other countries.

5. （略）

Section C

1. 1) T 2) F 3) F 4) T 5) F
2. 1) on the sea 2) the nomadic tribes 3) the splendid culture 4) Xuan Zang 5) VIP passport
3. 1) Silk Road was formally established during the Han dynasty of China, which linked the regions of the ancient world in commerce. The term for this network of roads was coined by Ferdinand von Richthofen, German geographer and traveler, in 1877 AD.

    2) The Silk Road reached its golden age in Tang dynasty, when the Chinese empire welcomed foreign cultures, making it very cosmopolitan in its urban centers. In addition to the land route, the maritime Silk Route was also developed. Chinese envoys had been sailing through the Indian Ocean to India since perhaps the 2nd century BC.

    3) After the fall of the Mongol Empire, the great political powers along the Silk Road became economically and culturally separated. The silk

trade continued to flourish until it was disrupted by the collapse of the Safavid Empire in the 1720s.

4. The famous Silk Road is a series of routes connecting the East and the West. The Silk Road extends more than 6,000 kilometers. It was named after the silk trade in ancient China. The trade in the Silk Road played an important role in the development of civilization in China, South Asia, Europe and the Middle East. It is through the Silk Road, Chinese papermaking, gunpowder, compass and printing of the four great inventions of ancient China were introduced around the world. Similarly, China's silk, tea and porcelain also spread all over the world. The exchange of material culture is two-way. Europe also exports all kinds of goods through the Silk Road to meet the needs of the Chinese market.

5. （略）

Section D

1. 1) F 2) T 3) F 4) T 5) T
2. 1) Silk Road Economic Belt, 21st Century Maritime Silk Road 2) the efficient allocation of resource 3) Peaceful Coexistence 4) people-to-people bonds 5) Indonesia
3. 1) The Belt and Road initiative refers to the Silk Road Economic Belt and 21st Century Maritime Silk Road, a significant development strategy launched by the Chinese government with the intention of promoting economic cooperation among countries along the proposed Belt and Road routes. The Initiative has been designed to enhance the orderly free-flow of economic factors and the efficient allocation of resources. It is also intended to further market integration and create a regional economic cooperation framework of benefit to all.

   2) Asian Infrastructure Investment Bank is a multilateral development bank that aims to support the building of infrastructure in the Asia-Pacific region, to promote the construction of regional connectivity and strengthen the cooperation of China and other Asian countries and regions. The bank is proposed as an initiative by the government of China and is headquartered in Beijing. Till 2017, it has 77 member states.

4. In the 21st century, a new era marked by the theme of peace, development, cooperation and mutual benefit, it is all the more important for us to carry on the Silk Road Spirit in face of the weak recovery of the global economy, the complex international and regional situations. When Chinese President Xi Jinping visited Central Asia and Southeast Asia in September and October of 2013, he raised the initiative of jointly building the Silk Road Economic Belt and the 21st-Century Maritime Silk Road, which have attracted close attention from all over the world. Accelerating the building of the Belt and Road can help promote the economic prosperity of the countries along the Belt and Road and regional economic cooperation, strengthen exchanges and mutual learning between different civilizations, and promote world peace and development. It is a great undertaking that will benefit people around the world.
5. （略）

Chapter 8
Section A

1. 1) T   2) F   3) F   4) T   5) T
2. 1) southwest   2) Lu Yu   3) Qing Dynasty   4) Green steaming   5) medicine
3. 1) It is said that tea was firstly discovered and tasted by Emperor Shennong. There is a famous legend of *Shennong Tasting A Hundred Plants*. One day, after walking for a long time, Shennong felt tired and thirsty, so he rested under a tree and started a fire to boil water in a pot. Suddenly some leaves fell into the pot from a nearby tree. Shennong drank the water and found it not only sweet and tasty but freshening as well. Another version said that Shennong tried 72 different kinds of poisonous plants in a day and he lay on the ground, barely alive. At this moment, he noticed several rather fragrant leaves dropping from the tree beside him. Out of curiosity and habit, Shennong put the leaves into his mouth and chewed them. After a little while he felt well and energetic again. So he picked more leaves to eat and thus detoxified his body from all the poisons. Those legendary leaves are said to be the earliest tea.

2) Tang Dynasty was a milestone in the development of tea industry in ancient China. In Tang Dynasty, a method of tea processing called "green steaming" was invented. Tang Dynasty's prominence was not only in the production and marketing of tea, but also the development of tea culture in this period. Tang Dynasty saw the first definitive publication about tea—*Cha Jing*, which was the first of its kind in the world. Its author, Lu Yu, was consequently dubbed the "Saint of Tea" by later generations. What's more, tea became the most popular commodity in foreign trades, and Japanese Buddhists brought tea seeds from China to Japan.

4. "Would you like tea or coffee?" This is the frequently asked questions at meals. Many westerners will choose coffee, while the Chinese will choose tea. It is said that a Chinese emperor discovered tea 5,000 years ago and used tea to heal illness. In the Ming and Qing Dynasties, teahouses spread all over the country. Tea drinking spread to Japan in the 6th century, but it was not until the 18th century that tea spread to Europe and the United States. Today, tea is one of the most popular beverages in the world. Tea is the treasure of China, and also an important component of Chinese tradition and culture.

5. （略）

Section B

1. 1) F 2) T 3) F 4) T 5) F
2. 1) green tea 2) Jiangsu 3) black tea 4) oolong tea 5) Dark tea
3. 1) Green tea is the oldest type of tea in China, and it is also produced in the greatest quantities. Many provinces and cities are renowned for their production of green tea, the most eminent provinces being Zhejiang, Jiangxi and Anhui. The leaves of green tea are not fermented, so they largely retain the original flavor of tea, which is simple, elegant and enduring. To make green tea, the methods used are mainly (green) steaming, (green) frying and (green) sun-drying, to remove the moisture in the fresh tea leaves and to bring out their fragrance. There are well-known varieties, such as West Lake Longjing Tea of Zhejiang Province, Biluochun Tea of Jiangsu Province, etc.

2) Oolong tea is a specialty from southeastern China, originating from provinces of Fujian, Guangdong and Taiwan. Oolong tea features a partial fermentation process, and thus has the characteristics of both green and black teas. Different varieties of oolong tea are processed differently, some are rolled into long curly leaves, while others are wrap-curled into small beads, each with a tail. The former style is the more traditional. Also, high quality oolong teas produce a long aftertaste that lingers in your mouth. In recent years it has been welcomed by more and more people with the name of "weight losing tea" and "bodybuilding tea". Tieguanyin, Dahongpao and Wuyi Rock Tea from Fujian as well as Dongding Oolong Tea from Taiwan are among the most prized oolong teas.

4. *Gongfu* tea is not one kind of tea or the name of tea, but a skill of brewing tea. People call it *Gongfu* tea for the reason of its exquisite process. The operational procedures require certain techniques, knowledge and skill of brewing and tasting tea. *Gongfu* tea originated in the Song Dynasty and prevailed mostly in Chaozhou, Guangdong Province (Chaoshan Area). It later became popular around the nation. *Gongfu* tea is famous for its high concentration. Oolong tea is mainly used in making *Gongfu* tea because it can meet the requirements of color, flavor and taste of *Gongfu* tea.

5. (略)

Section C

1. 1) T 2) T 3) F 4) F 5) T
2. 1) rude or offensive 2) The lower ranking person 3) the joining together of the bride and the groom's families 4) Teahouse 5) Tea ceremony
3. 1) Dating back to Tang Dynasty, *gongfu* tea ceremony is now the most famous type of Chinese tea ceremony and popular in Chaoshan area. The most important thing of *gongfu* tea is the tea sets. There are at least ten tea sets for Chaoshan *gongfu* tea ceremony. The way to practice *gongfu* tea ceremony include five basic steps. Step 1, prepare a bottle of boiled water. Step 2, put the tea leaves into the tea cup with hot water and soak for about 30 seconds, and then spill the water. Step 3, put the tea leaves into the tea funnel to filter out the impurities. Step 4, pour the

hot water again, and use the cup lid to stir the tea leaves a little bit. Step 5, pour the tea into the tea funnel again, and it is ready to drink.

2) Etiquettes in *gaiwan* tea ceremony include: first, serve tea with two hands holding the saucer and bow slightly forward. Make sure the guests don't have to move or stand up to receive the gaiwan. Second, those receiving the tea should hold the saucer but not the cup as the gaiwan cup itself can be hot. Third, they should hold the saucer to move the cup close to their mouths, which is the most traditional way of drinking tea from gaiwan. Finally, once they've drank the tea, take back the gaiwan with two hands, once again by holding the saucer.

4. China is a country with a time-honored civilization and also a land of ceremony and decorum. Whenever guests visit, it is necessary to make and serve tea to them. Before serving tea, you may ask them for their preferences as to what kind of tea they fancy, and serve them the tea in the most appropriate teacups. In the course of serving tea, the host should take careful note of how much water remains in the guests' cups. Usually, if the tea is made in a teacup, boiling water should be added into the cup when half of the tea in it has been consumed; And thus the cup is kept filled and the tea retains the same bouquet.

5. (略)

Chapter 9

Section A

1. 1) F 2) T 3) T 4) F 5) F
2. 1) millet 2) sugary substances 3) alcohol concentration 4) clear, tasty
   5) sugar, acidity
3. 1) As one of the six world-famous varieties of spirits (the other five being brandy, whisky, rum, vodka, and gin), *baijiu* is a distilled alcoholic drink with a relatively high alcohol content. It's made from various grains such as sorghum, millet, corn, rice and wheat. And it's renowned for being crystal clear, aromatic and tasty. In terms of its fragrance, *baijiu* is primarily classified into 6 categories: sauce fragrance, strong fragrance,

light fragrance, rice fragrance, phoenix fragrance and mixed fragrance. It is usually stored at least for several years to develop its full fragrance and flavor before it hits the market.

2) As one of the three world's ancient alcoholic drinks (the other two being beer and wine), *huangjiu*, or yellow wine, typically contains less than 20% alcohol by volume. It is also made from various grains such as rice, millet, wheat, sorghum and corn. And it's known for its transparent amber color and luster, balmy fragrance, and sweet and mild taste. Judging by sugar content, *huangjiu* primarily includes 5 categories: dry, semi-dry, semi-sweet, sweet and extra-sweet. It is usually warmed up to 60-70℃ before drinking. Such temperature can help give off its aroma. Jimo Rice Wine, Fujian Rice Wine and Shaoxing Rice Wine are well-known *huangjiu*.

4. *Maotai* liquor, along with Scotch whisky and French cognac brandy, is one of the three world famous distilled spirits. It is only produced in Maotai Town, Guizhou Province. Its unique taste benefits from the climate, soil and water quality of Maotai Town. In Qing Dynasty, *maotai* became the first Chinese liquor produced on a massive scale—170 tons a year. In 1915, it won a gold medal at the Panama-Pacific Exposition, the first time to have gained an international reputation. Two years after the founding of the People's Republic of China, *maotai* was designated as national liquor. Since then, *maotai* has been used at official banquets to entertain foreign heads of state and distinguished guests visiting China.

5. (略)

Section B

1. 1) T 2) F 3) F 4) T 5) T
2. 1) one, six 2) status, seniority 3) lower, respect 4) well-educated 5) grace
3. 1) As an integral part of alcohol culture, *jiuling* has a long history dating back to the Western Zhou Dynasty. It was, at that time, introduced to regulate people's drinking behavior in case of excessive drinking. But during the Warring States Period, *jiuling* evolved into a particular way to create more joyous atmosphere while drinking. And since Tang

Dynasty, it has become popular and generated diverse artistic forms. In terms of drinkers' social status and literacy, *jiuling* in ancient times was mainly divided into two types, literary and common. The former was complicated and catered to the taste of the well-educated. It covered a wide range of activities such as connecting idioms, composing couplets or verses. While the latter was simple and much enjoyed by the less-educated. It also covered a series of activities such as singing, telling stories or jokes. *Jiuling* was so appealing to both refined and popular tastes that some of its activities are still popular today such as dice throwing, drum-and-pass game and finger guessing game.

2) The finger guessing game is another popular game. It involves two players at a time. Both drinkers hold out their fingers simultaneously while shouting out a number from 1 to 10 respectively. If one of them shouts out the total number of fingers held up by both sides, he wins. The loser takes a drink or does a performance as a punishment. If both respond with the correct answer, then goes another round of game. Amazingly, they don't shout out the number directly but some words containing that number. These words mainly convey good wishes.

4. *Jiuling* is a kind of table manners to break the ice while Chinese are drinking. In the Western Zhou Dynasty, it was introduced to regulate people's drinking behavior in case of excessive drinking. But during the Warring States Period, *jiuling* evolved into a particular way to create more joyous atmosphere in drinking. And since Tang Dynasty, it has become popular and generated diverse artistic forms. In terms of drinkers' social status and literacy, *jiuling* in ancient times was mainly divided into two types, literary and common. Generally speaking, *Jiuling* was appealing to both refined and popular tastes.

5. （略）

Section C

1. 1) T 2) T 3) F 4) T 5) T
2. 1) state monopoly 2) Zhao Kuangyin 3) morale 4) literary creativity
   5) love, commitment
3. 1) Feast at Hong Gate is a popular story related to power struggle in the

guise of drinking. It is about two warlords Xiang Yu and Liu Bang in the late Qin Dynasty. Xiang Yu arranged a feast at Hong Gate, a place in Shaanxi Province today, attempting to kill his rival Liu Bang. After a few rounds of drinks, Xiang Zhuang, one of Xiang Yu's generals, was asked to perform a sword dance with the real purpose to assassinate Liu Bang. But Liu Bang saw through the whole scheme and eventually narrowly escaped.

   2) In the Spring and Autumn Period, Gou Jian, king of Yue State, planned to lead his army to attack the state of Wu. Before his riding off to war, his people presented good wine to him. To improve morale, Gou Jian ordered to pour it into a river and then drank with his soldiers upstream. As a result, the morale was greatly boosted and his army became invincible. This is the well-known Gou Jian Pouring Rice Wine to Reward Soldiers.

4. In the past of China, alcohol was considered as sacred liquid and used to worship the heaven, the earth and ancestors, having nothing to do with people's diet. Then it was only enjoyed by the royals and nobles for the state monopoly on its production and distribution. With the improvement of wine-making techniques and its production, it has already became a common drink and has been closely related to people's daily life. Since ancient times, literati have loved drinking. Alcohol has inspired their literary creativity. However, some literati drowned their worries or were indulged in alcohol to avoid misfortune in turbulent times.

5. （略）

Chapter 10

Section A

1. 1) F 2) T 3) F 4) T 5) F
2. 1) Yue, Xiang 2) red rice wine 3) Chuan cuisine 4) Guangdong
   5) Zhe, Chuan
3. 1) Sweet and Sour Carp is a famous traditional dish in Lu cuisine. It is said that the dish originated in the city of Jinan and later gradually spread in the whole country. Since the fish symbol is frequently used to symbolize

the wish for more in the sense of good luck, good fortune and long life. Sweet and Sour Carp is commonly served in wedding and birthday feasts as a main course to convey the good wishes.

2) Hot Pot, one of the most delicious dishes of Sichuan Province, is noted for its spicy, hot, fresh and fragrant taste. It creates a culture of getting together and enjoying good time. In Sichuan, from officials to common people, almost no one does not prefer Hot Pot and all families can make this course.

4. The eight regional cuisines, each with its own unique features, are the most influential and representative cuisines in China. Having been pursuing the unity of color, aroma, taste and shape, they are often described in the following vivid ways: Su and Zhe cuisine are often described as a graceful and elegant Southern beauty, Lu and Hui cuisine as a simple and robust Northern man, Min and Yue cuisine as a refined and romantic son of a wealthy family, Chuan and Xiang cuisine as a talented and versatile celebrity. The eight major cuisines reflect the essence of Chinese cooking technology and represent the peak of Chinese food culture.

5. （略）

Section B

1. 1) F 2) T 3) F 4) T 5) T
2. 1) *Jiaozi*, *niangao* 2) completeness, family reunion 3) sesame paste
   4) Huzhou, Jiaxing 5) Mid-Autumn Festival
3. 1) *Niangao*, another essential food for the Spring Festival, is very popular throughout China, especially in the South, because as a homophone, *Niangao* means "higher and higher, one year after another". *Niangao*, a New Year cake made of glutinous rice flour, is the necessary Spring Festival food for every family.

   2) The eight-treasure porridge, also called *Laba* Porridge, a special porridge eaten on the eighth day of the twelfth month, is made of glutinous rice, red beans, millet, Chinese sorghum, peanuts, dried lotus seeds, dried dates and almond or other nuts. The *Laba* Porridge was first introduced to China in the Song Dynasty about 900 years ago. In cold northern China, the eight-treasure porridge has become a

welcome food for breakfast since it is served hot and full of nutrition.
4. Chinese people value Mid-Autumn Festival for its important meaning of reunion. Moon cake is the traditional food for this special occasion. Traditional moon cakes have an imprint of the Chinese characters for "longevity" or "harmony", as well as the name of the bakery and the filling inside. There are many kinds of moon cakes in China, such as the moon cakes of Beijing-style, Cantonese-style, Suzhou-style, Yunnan-style and so on. Over time, both the skin and the fillings of moon cakes have diversified to cater to the changing taste. Traditional moon cakes are often sold alongside the new styles such as the Snowy Moon Cakes, the French-style moon cakes and so on.
5. （略）

参考文献

白东升，2005. 酒文化［M］. 呼和浩特：内蒙古人民出版社.
曹荣，华智亚，2006. 民间庙会［M］. 北京：中国社会出版社.
陈天水，1988. 中国古代神话［M］. 上海：上海古籍出版社.
陈永昊，余连祥，张传峰，1995. 中国丝绸文化［M］. 杭州：浙江摄影出版社.
邓杉，赵蓉，2010. 中西传统节庆文化概述［M］. 昆明：云南大学出版社.
丁以寿，2006. 中华茶艺［M］. 北京：中国农业出版社.
董俊峰，2012. 中华文化［M］. 长沙：湖南师范大学出版社.
方李莉，2005. 中国陶瓷［M］. 北京：五洲传播出版社.
方映，韩均，王立松，2011. 中国文化概览［M］. 天津：天津大学出版社.
郭松义，2011. 水利史话［M］. 北京：社会科学文献出版社.
何跃青，2013. 中国饮食文化［M］. 北京：外文出版社.
胡自山，等，2005. 中国饮食文化［M］. 北京：时事出版社.
蒋超，1994. 中国古代水利工程［M］. 北京：北京出版社.
金麦田，2004. 中国古代神话故事全集［M］. 北京：京华出版社.
刘勤晋，2006. 茶文化学［M］. 北京：中国农业出版社.
刘晔原，郑惠坚，1996. 中国古代的祭祀［M］. 北京：商务印刷馆国际有限公司.
龙毛忠，贾爱兵，颜静兰，2009. 中国文化概览（英文对照）［M］. 上海：华东理工大学出版社.
闵宗殿，纪曙春，1991. 中国农业文明史话［M］. 北京：中国广播电视出版社.
秦永洲，2015. 中国社会风俗史［M］. 武汉：武汉大学出版社.
宋镇豪，常玉芝，2010. 商代宗教祭祀［M］. 北京：中国社会科学出版社.
万建中，2004. 图文中国民俗·婚俗［M］. 北京：中国旅游出版社.
汪石满，2002. 中国民俗［M］. 合肥：安徽教育出版社.
王鲁地，2008. 中国酒文化欣赏［M］. 济南：山东大学出版社.
王渭泾，2009. 历览长河：黄河治理及其方略演变［M］. 郑州：黄河水利出版社.
王毓瑚，2006. 中国农学书录［M］. 上海：中华书局.
韦黎明，2005. 中国节日［M］. 北京：五洲传播出版社.
忻忠，陈锦，2009. 中国酒文化［M］. 济南：山东教育出版社.
徐昳荃，张克勤，赵荟菁，2016. 文化丝绸［M］. 苏州：苏州大学出版社.
阎万英，尹英华，1992. 中国农业发展史［M］. 天津：天津科学技术出版社.
杨敏，王克奇，王恒展，2006. 中国文化通览［M］. 北京：高等教育出版社.

游修龄，2014. 中国农耕文化漫谈［M］. 杭州：浙江大学出版社.

喻本伐，2016. 千年民俗文化［M］. 北京：清华大学出版社.

袁弟顺，2006. 中国白茶［M］. 厦门：厦门大学出版社.

赵佩霞，唐志强，2015. 中国农业文化精粹［M］. 北京：中国农业科学技术出版社.

周昕，1998. 中农具史纲及图谱［M］. 北京：中国建材工业出版社.

朱学西，1991. 五章中国古代著名水利工程［M］. 天津：天津教育出版社.

GLORIA K. FIERO, 2014. The Humanistic Tradition：From Prehistory to Medieval Period［M］. Beijing：Foreign Language Teaching and Research Press.

JONES A. Brief History of the Plough［DB/OL］. http：//www.ploughmen.co.uk/about-us/history-of-the-plough

LI ZHIYAN, CHENG QINHUA, 2002. Pottery and Porcelain［M］. Beijing：Foreign Language Press.

WHITFIELD S, SIMS-WILLIAMS U, 2004. The Silk Route：Trade, Travel, War and Faith［M］. London：British Library.

http：//baishi.baidu.com/watch/03295566731393874450.html

http：//edu.sina.com.cn/kids/2015-09-21/161292569.shtml

http：//en.chinaculture.org/library/2008-01/22/content_72460.htm

http：//en.chnmuseum.cn/english/tabid/549/Default.aspx?AntiqueLanguageID=2093

http：//english.anhuinews.com/system/2017/11/22/007755096.shtml

http：//english.cntv.cn/special/24_solar_terms/index.shtml

http：//english.gov.cn/beltAndRoad/

http：//english.gov.cn/news/top_news/2016/11/30/content_28147550424839htm

http：//english.visitbeijing.com.cn/a1/a-XCXWX762FEA9AB602B2A75

http：//ex.cssn.cn/sjxz/xsjdk/zgjd/zb/nj/fszs/

http：//factsanddetails.com/china/cat2/sub90/entry-5439.html

http：//factsanddetails.com/china/cat2/sub90/entry-5440.html

http：//factsanddetails.com/china/cat2/sub90/entry-5441.html

http：//factsanddetails.com/china/cat2/sub90/entry-5492.html

http：//factsanddetails.com/china/cat2/sub90/item50.html

http：//friends.unesco.hk/en/2017Apr/en/A4.html

http：//guji.artx.cn/article/8361.html.

https：//news.nationalgeographic.com/news/2008/01/080129-china-tomb.htm

http：//people.chinesecio.com/en/article/201003/23/content_115008.htm

http：//language.chinadaily.com.cn/2015-03-30/content_19950951.htm

http：//mygeologypage.ucdavis.edu/cowen/~gel115/115chxxyellow.html

http：//nancyjin.en.made-in-china.com/

http：//tcmwindow.com/cultivation/solarterms/Living-tips-at-Beginning-of-Spring.shtml http：//www.360doc.com/content/15/0107/14/6017453_438888000.shtml

http：//www.bxscience.edu/ourpages/auto/2013/10/16/60506571/yu%20the%20great.pdf

http：//www.china.org.cn/english/features/SpringFestival/200000.htm

http：//www.china.org.cn/learning_chinese/Chinese_tea/201108/02/content_23122957_2.htm

http：//www.china.org.cn/learning_chinese/Chinese_tea/201108/02/content_23122957_3.htm

http：//www.china.org.cn/opinion/2015-03/05/content_34962895.htm

http：//www.chinadaily.com.cn/beltandroadinitiative/index.html

http：//www.chinadaily.com.cn/cndy/2016-09/28/content_26921359.htm

http：//www.chinadaily.com.cn/culture/2016-02/04/content_23390783.htm

http：//www.chinaknowledge.de/History/Terms/wanganshireforms.html

http：//www.chinaknowledge.de/Literature/literature.html

http：//www.chinatoday.com.cn/english/culture/2017-06/09/content_741904

http：//www.chinatownplus.com/chinahand/rain-water-yu-shui-the-twenty-four-solar-terms/

http：//www.chinatraveldepot.com/C173-Chinese-Alcohol

http：//www.cits.net/china-travel-guide/chinese-24-solar-terms.html

http：//www.comuseum.com/ceramics/ming/

http：//www.comuseum.com/ceramics/qing/

http：//www.comuseum.com/ceramics/tang/

http：//www.ebeijing.gov.cn/feature_2/Traditional_Fesitival/Qingming_Festival/Legend_QF/t1021380.htm

http：//www.en8848.com.cn/read/story/zgmythology/155196.html

http：//www.foreignercn.com/index.php?option=com_content&view=article&id=5175：temple-fair-in-beijing&catid=1：history-and-culture&Itemid=114

http：//www.globaltimes.cn/content/1022645.shtml

http：//www.globaltimes.cn/content/1068304.shtml

http：//www.gov.cn/fwxx/wy/2008-01/15/content_858085.htm

http：//www.kekenet.com/read/201506/378846.shtml

http：//www.religionfacts.com/chinese-religion/veneration-ancestors

http：//www.sfcca.sg/en/node/62

http：//www.telegraph.co.uk/news/world/china-watch/culture/24-solar-terms/

http：//www.xinhuanet.com/english/china/2016-11/29/c_135866212_3.htm

https：//baike.baidu.com/item/%E5%B7%9D%E8%8F%9C/26409

https：//baike.baidu.com/item/%E5%BE%BD%E8%8F%9C

https：//baike.baidu.com/item/%E6%B5%99%E8%8F%9C

https://baike.baidu.com/item/%E6%B9%98%E8%8F%9C

https://baike.baidu.com/item/%E7%B2%A4%E8%8F%9C

https://baike.baidu.com/item/%E8%8B%8F%E8%8F%9C

https://baike.baidu.com/item/%E9%97%BD%E8%8F%9C

https://baike.baidu.com/item/%E9%B2%81%E8%8F%9C/458122?fr=aladdin

https://chinese-tea.net/

https://chinese-tea.net/#tab-con-1

https://chinese-tea.net/#tab-con-6

https://en.wikipedia.org/wiki/Alcoholic_drink

https://en.wikipedia.org/wiki/Alcoholic_drinks_in_China

https://en.wikipedia.org/wiki/Ancestor_veneration_in_China

https://en.wikipedia.org/wiki/Baijiu

https://en.wikipedia.org/wiki/Beer

https://en.wikipedia.org/wiki/Beer_in_China

https://en.wikipedia.org/wiki/Buddha_Jumps_Over_the_Wall

https://en.wikipedia.org/wiki/Chinese_ceramics

https://en.wikipedia.org/wiki/Chinese_funeral_rituals

https://en.wikipedia.org/wiki/Chinese_mythology

https://en.wikipedia.org/wiki/Customs_and_etiquette_in_Chinese_dining

https://en.wikipedia.org/wiki/Drink

https://en.wikipedia.org/wiki/Drinking_game

https://en.wikipedia.org/wiki/Huangjiu

https://en.wikipedia.org/wiki/Jingwei

https://en.wikipedia.org/wiki/Kua_Fu

https://en.wikipedia.org/wiki/Li_Bai

https://en.wikipedia.org/wiki/Miaohui

https://en.wikipedia.org/wiki/Mince_pie

https://en.wikipedia.org/wiki/Mooncake

https://en.wikipedia.org/wiki/Moxibustion

https://en.wikipedia.org/wiki/Myth

https://en.wikipedia.org/wiki/One_Belt_One_Road_Initiative

https://en.wikipedia.org/wiki/Pangu

https://en.wikipedia.org/wiki/Sacrifice

https://en.wikipedia.org/wiki/Shang_Yang

https://en.wikipedia.org/wiki/Shennong#Mythology

https://en.wikipedia.org/wiki/Silk_industry_in_China

https://en.wikipedia.org/wiki/Su_Shi

https：//en. wikipedia. org/wiki/Traditional _ Chinese _ holidays

https：//en. wikipedia. org/wiki/Wang _ Anshi♯Major _ reform

https：//en. wikipedia. org/wiki/Wine

https：//en. wikipedia. org/wiki/Wine _ in _ China

https：//en. wikipedia. org/wiki/Zongzi

https：//wenku. baidu. com/view/45725591910ef12d2bf9e765. html

https：//wiki. samurai-archives. com/index. php? title＝Single _ Whip _ Reform

https：//www. acufinder. com/Acupuncture ＋ Information/Detail/The ＋ History ＋ of ＋ Acupuncture

https：//www. chinaeducationaltours. com/guide/culture-solar-terms-in-autumn. htm

https：//www. chinahighlights. com/festivals/the-24-solar-terms. htm

https：//www. chinatravel. com/facts/ancient-rituals. htm

https：//www. chinatravel. com/facts/chinese-tea. htm

https：//www. chinatravel. com/facts/chinese-tea. htm

https：//www. chinatravel. com/facts/chinese-tea-categories. htm

https：//www. chinatravel. com/facts/chinese-tea-history. htm

https：//www. chinatravel. com/facts/chinese-wine. htm

https：//www. chinausfocus. com/finance-economy/one-belt-and-one-road-far-reaching-initiative/

https：//www. echineselearning. com/blog/learn-chinese-culture-24-solar-terms

https：//www. shine. cn/archive/sunday/now-and-then/The-24-Solar-Terms/shdaily. shtml

https：//www. theknot. com/content/ancient-chinese-wedding-tradition

https：//www. travelchinaguide. com/intro/cuisine _ drink/alcohol/

https：//www. travelchinaguide. com/intro/focus/solar-term. htm

https：//www. tripsavvy. com/how-to-say-cheers-in-chinese-1458379

https：//www. yourchineseastrology. com/calendar/24-solar-terms. htm

图书在版编目（CIP）数据

中国农业文化概览：英文 / 王珍主编 . —北京：中国农业出版社，2019.12（2024.2 重印）
普通高等教育农业农村部"十三五"规划教材　全国高等农林院校"十三五"规划教材
ISBN 978-7-109-26267-6

Ⅰ. ①中… Ⅱ. ①王… Ⅲ. ①农业史－文化史－中国－高等学校－教材－英文 Ⅳ. ①S-092

中国版本图书馆 CIP 数据核字（2019）第 284299 号

中国农业文化概览
ZHONGGUO NONGYE WENHUA GAILAN

中国农业出版社出版
地址：北京市朝阳区麦子店街 18 号楼
邮编：100125
责任编辑：马颋晨　李　想　　文字编辑：李　想
版式设计：王　晨　　责任校对：吴丽婷
印刷：北京通州皇家印刷厂
版次：2019 年 12 月第 1 版
印次：2024 年 2 月北京第 3 次印刷
发行：新华书店北京发行所
开本：720mm×960mm　1/16
印张：19.75
字数：345 千字
定价：39.50 元

版权所有·侵权必究
凡购买本社图书，如有印装质量问题，我社负责调换。
服务电话：010-59195115　010-59194918